LEEDS POLYTECHNIC - BECKETT PARK LIBRARY

Practical
Data
Analysis

P M Downes

BLENHEIM
ONLINE PUBLICATIONS
1989

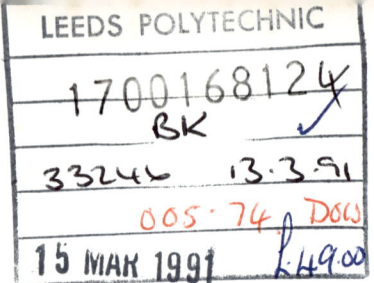

British Library Cataloguing in Publication Data

Downes, Patric
 Practical data analysis
 1. Machine-readable files. Design
 I. Title
 005.74

ISBN 0-86353-162-8

© Blenheim Online 1989

Whilst every effort has been made to ensure the accuracy of the information contained in this book, the Publisher cannot be held liable for any errors or omissions however caused.

The Publisher acknowledges the assistance of Spicer & Oppenheim Consultants in the production of the many illustrations in this book.

No part of this book may be reproduced, stored in any form, by any means, electronic, mechanical, photocopying, microfilming, recording or otherwise without written permission from the Publisher.

*Printed in the United Kingdom
by Hobbs the Printers Ltd, Southampton.*

Blenheim Online Publications

Blenheim House, Ash Hill Drive, Pinner, Middlesex HA5 2AE, UK

The Publishing Division of Blenheim Online Ltd, London

"The Universe contains 'Things'.

Things have 'Attributes'.

One Thing may have

many Attributes; and one

Attribute

may belong to many things."

Lewis Carroll
Symbolic Logic : Book 1, Chapter 1

PREFACE

This book has taken shape over a considerable period of time and its development has been assisted by many people. Certain individuals deserve special mention and my heartfelt thanks. In particular:

>Dr T W Olle, for increasing my own understanding;

>David J Salway, for encouragement and guidance;

>Richard J Armitage, for advice, general comment and review of the text;

>Carol Adams, Ania Piekos and Tony Freshwater, for text and graphics production;

>Rollo Turner, for guidance and advice on publication details; and

>Mary, for support, encouragement and patience throughout the whole process.

Several experts have been instrumental in the establishment of data analysis as an effective approach to system design. In particular I would like to acknowledge the work of Edward Codd, Charles Bachman and James Martin, who pioneered and promoted many of these techniques.

P M Downes

April 1989

CONTENTS

	Page
Introduction	xi

Part 1: Concepts and techniques

Chapter 1:	Why Analyse Data	
	Classical Analysis	3
	Database Rationale	7
	Value of Data	10
	Stability of Data	11
Chapter 2:	Data Analysis Concepts	
	Introduction to the 'EAR' Approach	13
	Entity Types	16
	Attributes	18
	Relationships	21
Chapter 3:	Basic Data Analysis Techniques	
	Concentrate on Business Data	29
	Time Dependencey	31
	Categorisation of Entities	35
	Events	41
	Modelling Hierarchies	44
	Recursive Structures	46
	Company Organisations	50
Chapter 4:	Techniques for Refining the Model	
	Entity Splitting	55
	Entity Combination	63
	Derived Data	66
	Coding Structures	68
	Cross-Reference Entities	74
	Role Entities	77
	Tightening Constraints	80
	Problems with Updates	84
Chapter 5:	Checking the Model	
	Establishing Identifiers	87
	Supportive Attributes	89
	Alternate Keys	95
	Normalisation	96
	Abstraction	101

Part 2: Using data analysis

Chapter 6: The Data Model

Diagram Conventions	109
Spotting Problems	111
Numbering the Entity Types	114
Quick and Dirty Models	116

Chapter 7: Back-up Documentation

Purpose of the Documentation	119
Content Overview	121
The Description	123
Examples	124
Volume	125
Growth Rates	127
Retention	128
Average Population	130
Source of Data	132
Maintenance Responsibility	133
Questions	134

Chapter 8: Developing a Data Model

Overview	135
How Often?	137
Controlling Re-Issues	140
Outputs	142
Checking Consistency Between Issues	144

Chapter 9: Using Existing Systems

Useful Sources	147
Using Existing Record Structures	149
Implementation Dependent Fields	152
Using Program Code	154
Exceptions Processing and Short-cut Solutions	156
Problems with these Sources	158

Chapter 10: Involving Users

Get Commitment	161
Establish Contacts	163
Talking to Users	165
Single User Interviews	168
Group Discussions	170
Establish Data Usage	172
Resolving Conflicts	174

Chapter 11: Coping with a Complex Model

The Problems	177
"Break It Down"	179
Cross Check the Sub Models	183

	Sub-Modelling Conventions	185
	Dangers and Their Avoidance	188

Part 3: Process analysis

Chapter 12: Analysing Processes

Introduction	193
Operations	197
Events	198
Levels of Nesting	200
Other Useful Data	208

Chapter 13: General Techniques for Process Analysis

Types of Process	211
The Purpose of the Process	213
The Nature of the Process	217
People and Positions	221
Implementation Constraints	222
Historical Processes	224
Errors and Omissions	225

Chapter 14: Combining Process and Data Models

Introduction to Process Mapping	227
Techniques	231
Noting Access Points	235
CRUD	237
Other Process Detail	240
Checking the Results	241

Part 4: Methodology

Chapter 15: The Method Described

Data Analysis in Context	245
Defining the Scope	248
Establishing Control	250
Initial Information	253
The First Model	255
Reviewing the Model	258
Iteration	260
Reviews and Talk-throughs	263
Get It Agreed	264
Overall Review	265
Method Summary and Text Cross-Reference	266

Part 5: Using the model

Chapter 16: Changes to the Existing Business

Introduction	271

	How to Use the Model	272
	Don't Short Cut	274
Chapter 17:	New Business Areas	
	Extending the Model	277
	Revise Existing Structure	279
	Integration	283
Chapter 18:	Going Database	
	The Story so Far	285
	Getting Started	289
	Which Database?	291
	Phasing the Implementation	300
	Database Design	304
	Entities vs Records	305
	Attributes vs Fields	308
	Relationships vs Sets and Indexes	310
Chapter 19:	Physical Design Issues	
	Performance	313
	Security and Recovery	317
	Storage	319
	Redundancy	321
	Reorganisation	323
	Housekeeping	325
Chapter 20:	Other Related Issues	
	User Staff	329
	Systems and Programming Staff	331
	Data Management and Database Administration	333
	Data Dictionaries	335
	Query Languages	339
	Other Data Management Software	341
Chapter 21:	The Way Forward	
	Why Use Data Analysis?	343
	DBMS or not DBMS?	345
	Future Scenarios	346
	In Conclusion	348

Appendix

Sample documentation

Introduction	351
Data Analysis Documentation	352
Process Analysis Documentation	375

INTRODUCTION

Why 'Practical' Data Analysis?

The aim in writing this book is to provide detailed and practical advice on how to organise and carry through a successful data analysis project. In effect, it is aimed primarily at answering the question 'How do I analyse data?'. This book does not set out to address Relational Theory or any of the 'academic' issues underlying the techniques proposed, except where a knowledge of these is essential to their practical usage. The debates concerning 'Why analyse data?' and, to a lesser extent, 'What is data analysis?', have been covered by experts of international repute and the arguments are well known. But it still seems rare to find a text that actually tells the reader *how* to carry out data analysis, rather than *what* to do. The intention of this book is to fill that gap.

Each of the proposed techniques and points discussed in the following pages has been used by the author and each is included only because it has been found to be of practical benefit in achieving a well documented data model. There are other techniques available which are not discussed here. This is usually due to one of two reasons. Either the technique has not been found to be beneficial in practice, or it adds no new knowledge over and above that generated by other, often simpler, approaches. Where two techniques are roughly equivalent, this text always proposes the simpler approach thus minimising complexity and maximising ease of use.

The reader will find very few hard and fast 'rules' expressed in the following pages. This is simply due to the fact that in the real world, when faced with real problems, there is often more than one right answer, the best choice depending on the precise circumstances at the time.

Hence the emphasis is on guidelines rather than rules, although wherever possible the likely practical alternatives are described together with discussion of the factors affecting the choice between them.

This book is written for use by data processing staff who wish to familiarise themselves quickly with what data analysis is in practice and how it may be used. Alternatively, the book may be used as a reference point or check-list for data analysis practitioners who wish to ensure that all major facets of a data modelling exercise have been addressed.

Structure of the Book

This book is broken down into five parts. The first part, Concepts and Techniques, begins with an introduction to data analysis terminology and data modelling conventions. It examines the reasons why data analysis has risen to prominence and the benefits that can accrue from use of the techniques. The main data analysis techniques are described at length, along with examples, when and how each should be used, and any practical implications of using each technique.

Part 2, Using data analysis, considers the documentation of the results of data anlysis and some further data modelling conventions for use on large and complex projects. Sources of information and the best ways of utilising these are also discussed along with an explanation of the iterative nature of data analysis and the difficulties that can arise. Considerable details on involving users is given in this part, as consistent and full user involvement is an essential ingredient in successful modelling.

Process analysis, in some ways the opposite side of the coin to data analysis, is discussed in Part 3. The interplay between the two is examined and the technique of using process analysis to cross-check a data model is explained. Documentation standards for process analysis and also for the results of the cross-checking, are proposed.

Part 4 attempts to pull the preceding three sections together to form a 'Basic Methodology' which is flexible and can be shaped to individual requirements. Each stage in the method is described and this is followed with a tabular presentation (*Figure 15.5*) of each step with a cross-reference to the appropriate sections in the text. It is intended that this detailed method may be 'slotted in' to any existing methodology which the reader may use. The method proposed is not put forward as a complete system design methodology, but addresses only the Logical Analysis/Design phase, culminating in the production of a fully documented data and process model of the business functions.

The final part examines some of the ways in which a data model may be used to help other data processing tasks. It also includes an introduction to database design and the types of decisions that need to be made when designing a database from a data model. Most sections of each chapter start with a summary on the points covered in that section. These summaries enable easier access by providing a rapid reminder to the full content.

An Appendix containing sample documentation has been included. This gives an insight into the types of documentation that one would expect to see although it is not a complete set as this would be unwieldy in a book of this size. The documentation covers both process and data analysis outputs and their interrelationships.

Although data analysis does not imply a database solution, data analysis and share data becomes much more important. The complexity of these systems is also of note. Information is typically used in a variety of ways and the volume of information involved necessitates some automation of these processes. It is the flexibility inherent in database software that allows this complexity to be supported. Additionally, conventional approaches to system design cause difficulty where information needs to be shared across a number of applications and in a variety of ways. However, these 'advantages' of sharability of information and flexibility of the software can very rapidly turn into disadvantages if the data is poorly structured; hence the need for data analysis.

If a database solution is to be adopted, then the control and a management of the data becomes very important. In these situations it is essential that a thorough data analysis exercise is carried out prior to database design and preferably prior to software selection.

Part 1

Concepts and Techniques

Chapter 1

Why Analyse Data?

Classical Analysis

Classical analysis has typically led to the following problems:

- *complex update routines;*
- *inconsistent data;*
- *data duplication;*
- *data redundancy;*
- *inflexibility.*

Traditional approaches to systems design have always taken as their primary concern the processes that the system is to carry out. Any consideration of the data requirements and the structuring of that data has always been from the viewpoint of the application itself. In short, data has been seen as a means to an end. This is, of course, something of an exaggeration as some data design has always occurred as a by-product of system design, but it is the underlying philosophy of 'processes before data' that characterises classical approaches to systems design.

This emphasis on 'processes before data' leads to various problems, most of which can be found in any traditional data processing installation. This can be summarised as an unhealthy degree of systems independence. A typical scenario is given below.

> The A N Other Manufacturing Company has a number of systems running, all of which have been developed in-house over a period of some years. Firstly, there is the payroll system developed several years ago, running in batch mode and with a weekly update. Secondly, the production system which records details of the day's production and is updated every night. Lastly, there is a brand-new, fine and dandy resource scheduling system which is on-line with real-time main file updating.

This is shown in Figure 1.1. In many installations it is not unusual for there to be no communication between these systems and each has its own discrete data store.

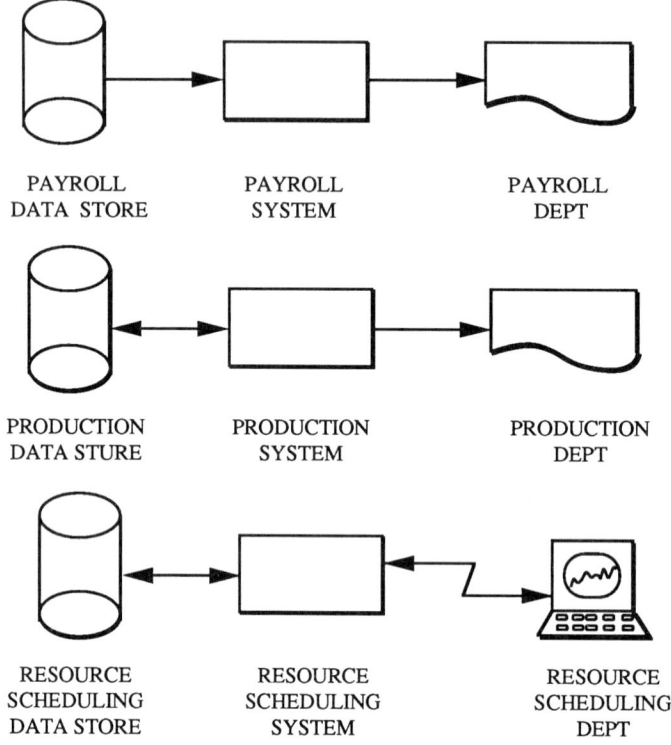

Figure 1.1 A typical installation scenario

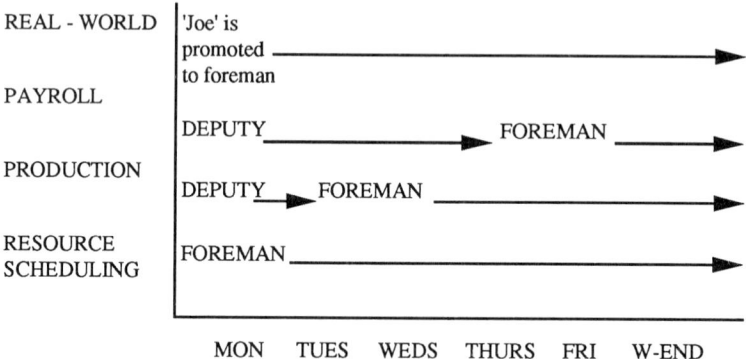

Figure 1.2 Due to different update cycles, different systems show different information

Assuming for convenience that all three systems contain details on individual employees, we start to get some idea of the type of problems that will arise. If the details concerning an employee change, let us assume they get promoted, then all three systems have to be updated with the revised data. With conventional approaches there are two basic ways to achieve this. Firstly, a set of manual procedures can be introduced to ensure that all three systems, probably developed for different user departments, are updated. These procedures will almost certainly be fairly complex and difficult to control and it is extremely likely that failures will occur. At the very least there are now three input stages and the chances of miskeying have been multiplied by the same factor.

If this is felt to be unsatisfactory then the second approach is to attempt to make the transaction files from one system feed the others and to automate the propagation of the change without manual control. The likelihood here, though, is that each system will hold the data in differing formats or that different coding structures will be used. The overall result is that the programs dealing with the transfer of the data will be unwieldy and extensive re-formatting of the data will have to be done in order to turn one system's data into another.

So what are the end results of these techniques? Firstly, as indicated above, update problems and difficulties immediately emerge. They are either difficult to administer or time-consuming to maintain when changes occur. Secondly, data can easily become inconsistent across these three systems if failures in the updating process occur, and this in itself can cause problems which may be difficult to trace and can remain hidden for a considerable length of time.

Even where all the controls and procedures work correctly we can still encounter difficulties. In the example outlined above, and as shown in Figure 1.2, the three systems all have different update cycles. This means that for some period of time the data held by the systems will be inconsistent: if enquiries are made the results will vary depending on which system is interrogated.

In addition, the same or similar information is held three times, duplicating information unnecessarily and wasting storage space. In large organisations where this typical pattern is repeated across many systems and for many data occurrences, the problems multiply and the cost of wasted storage alone can become significant. Also, in these situations, redundant data is often held and duplicated across many files, 'in case it is needed somewhere else' being the standard thinking. Redundant data is, again, a waste of resources and can clutter a system with unnecessary and out-of-date information.

The final problem is inflexibility for enquiries. If a simple question needs to be asked of a number of systems, then each enquiry has to be formatted differently for each system and the answers will be obtained in different ways and at different times. These results then have to be collated and interpreted to give the total picture. This again is assuming that the data held is consistent and up-to-date on all systems. If it is contradictory then the problems multiply once more.

How does data analysis overcome these problems? The starting point can be called the Database Rationale.

Database Rationale

Using the database rationale helps to achieve the following:
- *less data duplication;*
- *sharable data;*
- *consistent data;*
- *control over data validation;*
- *better enquiries;*
- *data independence from process changes;*
- *better security and recovery of data files.*

Many of the problems that have been discussed are not the result of poor systems analysis or faulty design. Fundamentally, the cause has been the inability of conventional computer files to represent data in a natural and structured way. Conventional flat-files are best suited to relatively large records but relatively few record types, whereas in the real world data is much more diverse and consists normally of fairly small chunks. Recent developments in database technology have resulted in reliable software products and file handling techniques that can represent data in a much more natural way.

Although data analysis is not necessarily associated with database software, the two have become closely related as data analysis itself results in a structured representation of data where the 'records' tend to be small but the number of 'record types' fairly large. This is a general approach which may be termed the Database Rationale and it does address the problems that we have already discussed.

The main thrust of this approach is that it is the data that should be examined first with the processes that use the data being examined later in the design process. This is not to say that the processes are not important but rather that, if the data can be broken down and structured in a very real sense, then the processes that use the data will be simplified and more easily maintained. In a sense then, data analysis attempts to put the cart before the horse, stressing the need to establish the data requirements and the relationships between data elements (the cart) and then to analyse the processes that use the data (the horse).

Part of this argument assumes from the outset that data should exist once only and should be sharable. If a customer, for example, is involved with an organisation in, let us say, five different ways, then that customer data should only be held once but made available for use by all five processes. This is an inherently more natural approach to computer systems design. The customer only exists once in the outside world and, given that the aim is to produce systems which help an organisation function in that world, the more natural the representation the better.

By adopting this approach a number of problems have been solved immediately. We have achieved our economy-of-storage objective as information is no longer needlessly duplicated across systems. The commonality and sharability also ensure that the stored data is much more consistent and less contradictory. In the earlier example of the A N Other Manufacturing Company, if employee details had been stored only once then the change would only have to be applied once for all three systems to utilise the update. Enquiries would always yield consistent results with no need either for complex file re-formatting programs or for involved manual procedures to control changes. Additionally, where information is mis-keyed or incorrectly updated, the error is much more noticeable and easier to correct.

Another benefit of this approach, and one that is often under-appreciated, concerns manipulation of rules and data vetting. It is very common for computers to get the blame for all kinds of silly mistakes and other errors. We may all be aware of such phrases as 'Garbage In, Garbage Out' (GIGO) and 'Keep it Simple, Stupid' (KISS) but what do these phrases really mean?

Unfortunately it is very difficult to keep things simple. We live in a complex world where situations do change and a lot of flexibility and change is required. Also, the reason that garbage gets into a system in the first place is that the rules governing input to a system are normally held within program code, and have been developed from a very narrow system-orientated point of view: for example if Field A has value (x), then Field B must be in the range 1-100. These rules are also very difficult to change, especially when they re-occur in several programs. Each needs to be amended and re-tested and each exception to the general rule has to be explicitly stated in the program code.

Data analysis provides the ability to structure these rules and take them out of the program code. Rules are data, albeit of a slightly different kind. It is, if you like, data governing data and, as we shall see in later chapters, this data can be modelled and structured within the data model. Again, when the same rules re-occur several times, they only need to be held once as these, like any other data, can be shared and kept consistent. Once the mental leap to treating rules as data has been made, it is then but a short step to *allow* these rules to be manipulated and updated directly by the users thus allowing easier maintenance and lightening the load on hard pressed DP departments.

To give maximum benefit, data analysis needs to be an organisation-wide approach. It is possible to take a small part of the organisation and draw up a data model. The technique will work and the benefits will accrue, but many of the goals of sharability and data relationships will not be attained. A corporate data model can be a difficult thing to achieve in practice as many organisations do not have the time or the resources to embark upon an all-singing, all-dancing exercise of this size. However, all is not lost; there are ways of working towards this goal which although not ideal, will allow gradual development of a corporate model by utilising a phased approach to the problem without requiring too much reworking.

The ability to achieve better enquiry facilities cannot be overstressed. There is an increasing movement in recent years away from an emphasis on traditional 'data processing' and an increasing requirement for 'information processing'. The need for management to have accurate up-to-date information available to them is becoming more and more important. The main difficulty in this area is the variability of the enquiries that need to be satisfied. To achieve this more powerful query languages are needed and a clearly understood and wide-ranging data structure is necessary. Databases, with their in-built flexibility, offer a realistic means of attacking some of these difficulties.

A further area of traditional concern has been over assessing the impact, and controlling the implementation of changes to the processing requirements as the business changes. As the data has been designed to fit the process, if the process changes then the data will be affected as well. By treating the data as something much more independent of the process from the outset then changes to processing requirements tend to be less traumatic. Even in those situations where changes to data structures are required then the impact of these can be more easily assessed and planned when all the relationships within the data structure have been identified.

Lastly, and perhaps more as a function of database software rather than data analysis, security and recovery of the data can be tackled in a more co-ordinated and uniform way. Again, this is due to the philosophical leap. Data no longer 'belongs' to the system, but is merely *used by* a system. Data security can then be handled consistently and many database packages provide useful and effective security and recovery facilities.

Value of Data

Data belongs to the whole organisation, not just a part of it. As a result data must be managed and controlled. Database administration becomes an essential function.

Awareness of the value of data has dawned relatively slowly in the corporate minds of many organisations. Data is as valuable a resource as any other asset of an organisation such as buildings or personnel skills. If data is lost the consequences for even a small company can be quite catastrophic. The cost of recovering from this situation in terms of lost business, non-availability of information, etc, can be crippling. Although computer departments have taken some steps to prevent loss of data, this has again normally been treated as a somewhat low-profile overhead rather than as an essential part of computer strategy. As the quantity of data expands and reliance on computer systems grows, then this area becomes more prominent and needs a co-ordinated and controlled strategy in order to make it effective. Although data analysis does not address this directly, it does promote an awareness of data and, as mentioned earlier, security strategies are one of the spin-off benefits of database technology.

Data belongs to an organisation rather than to a system or one department. In this sense it is necessary to control and manage data from a global point of view. If data is going to be shared and used in many parts of the organisation then that data requires central management independent of any single application area. If data is not managed and controlled, and user departments are allowed to manipulate and change it for their own requirements alone, then we are back to all the problems that have been outlined earlier. Although data analysis allows control of data to be put back to the users who need it, that responsibility requires management and co-ordination in order to be successful.

A powerful database administration function becomes essential and the responsibilities and skills needed in this area are discussed towards the end of this book.

Stability of Data

Conventional systems under-utilise the stability of data and often ignore certain types of data. This leads to:

- *poor use of historical data;*
- *inadequate data vetting;*
- *difficulty in interpreting aggregations and summaries.*

Most data, and in particular data structures, are unchanging. *Data exists independently of the processes which use it.* In most cases, and in absolute terms, data is not changed by processes; rather, data is added or taken away. The concepts of 'What has been' and 'What will be' have not been considered in any depth by most computer applications. The result of this is that data on conventional systems appears to be quite volatile. As data is added, old values are often thrown away and the contents of files appear to be ever-changing. This is due to conventional systems concentrating on two aspects of the same problem. Firstly, the emphasis is again on the processes and what they do rather than the framework within which they operate. Secondly, there is a lot of emphasis on the current situation and a disregard of past or future situations.

This book takes the philosophy that there is much data of interest which is rarely, if ever, found in computer systems. Organisation structures and the way in which the organisation views the outside world and its position in it are often totally ignored. This information is essential if the data input to the system is to be fully validated on input and the many rules and relationships are to be utilised – lack of attention to this is often a cause of the GIGO problem. This data, once established, rarely changes in structure and the formalisation of the data analysis process itself causes problems in this area to be highlighted and corrective action taken. This data is also of use if enquiries are to be interpreted correctly. Summaries and aggregations of data need to be produced in terms of these structures, whether it be by departmental responsibility, customer groupings or whatever.

The last point is that computerised information is only a representation of something that exists in the outside world. This data exists whether or not there is a computer system to use it. In a very real sense then, the data has an independent existence and is not transient or system-dependent.

This shift in philosophical stance is necessary if a complete understanding of the value of data analysis is to be gained. Data analysis is not just a collection of techniques but is a different way of looking at computer applications and the way in which they should be developed. If the underlying philosophy can be grasped, then the concepts employed and the techniques used will be seen to fit together into a co-ordinated and thorough approach.

Chapter 2

Data Analysis Concepts

Introduction to the 'EAR' Approach

Data analysis employs three major concepts:

- *Entity type: a collection of data elements which forms an identifiable unit;*
- *Attribute: an individual data element;*
- *Relationship: denoting an interdependency between two entity types.*

Although the previous chapter discussed the philosophy of data analysis, it did not specifically attempt to define it. There are a number of equally valid practical approaches to data analysis and all of them fit with a general definition of 'a collection of techniques to establish the data important to an organisation and to identify the structure of that data and the relationships between different data elements'.

The aim, therefore, is to establish a data model showing the structure of the data that is relevant to the organisation, and some indication of the contents of that structure. It is at this stage where we need to introduce some jargon in order to establish clearly what it is we are discussing. The approach taken in this book, takes its name from the three bits of jargon used, these being entities, attributes and relationships (EAR). Any reader familiar with database techniques will have noticed a close correspondence between entities and records, attributes and data items, and relationships and data sets. It is important, though, to use the correct terminology so that we know when we are talking about business data (entities, attributes, relationships) as opposed to a computer implementation (records, data items, data sets). Data analysis does not necessarily imply a computer-based solution, and it is essential that the two are kept separate.

An entity, or more correctly, an *entity type*, is a collection of data elements that when taken together form an identifiable unit. Customer, for example, is a likely candidate for an entity type and would be made up of various data elements or bits of information such as Customer Number, Customer Name, Telephone Number etc. An example of the Customer entity type might be, for example the A N Other Manufacturing Company, this being a particular occurrence of the entity type. *Attributes* are those bits of data that go to make up an entity type. Therefore, the entity type Customer, would have the attributes Customer Number, Customer Name, etc. Finally, a *relationship* describes the interdependence of one entity type upon another. For example, there may be two entity types, Customer and Order, and there would almost certainly be a relationship between these two. Customers place Orders so the Order must be related to an associated Customer.

As pointed out earlier, the aim is to produce a data model which is a representation of the data structure in diagrammatic form. So it is necessary to establish a convention for this diagram or data model. Several such conventions are available but possibly the most straight-forward is the one used within this book.

Entity types are represented as shown in Figure 2.1. As can be seen, this is nothing more than a box with a title naming the entity type. In a data model, the entity type CUSTOMER would only be shown once, even though there will be many occurrences of this entity type. The use of block capitals when referring to entity types is a convenience, and will be used in the remainder of this text. An entity type is always named on the model in the singular: there is only one entity type CUSTOMER representing many customers – the entity occurrences.

```
┌─────────────────┐
│                 │
│    CUSTOMER     │
│                 │
└─────────────────┘
```

Figure 2.1   Representation of Entity Types

Attributes can be shown as a simple list for each entity type as shown in Figure 2.2. Avoid the use of block capitals when naming attributes, as this can be confused with entity types. Similarly abbreviations should be avoided as these can be misinterpreted by different readers. Attribute lists are useful as working documents although they do not form part of the final documentation and should be held separately from the data model.

CUSTOMER: [Customer Number, Customer Name, Address, Telephone Number, Credit Limit etc]

Figure 2.2   Indicating the Attributes of an Entity Type

Relationships are shown as a line connecting two entity types as seen in Figure 2.3. There are several different types of relationship and each is represented in a slightly different way. Each type, and the significance of the arrow on the line representing the relationship, will be discussed later in this chapter.

It is now appropriate to consider entity types, attributes and relationships in more detail.

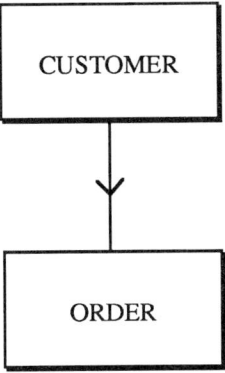

Figure 2.3   Representation of a relationship between the entity types CUSTOMER and ORDER

Entity Types

An initial list of entity types can be discovered by identifying the nouns used to describe the business. However, the following points should be considered:

- *is the concept 'real'?*
- *create entity types in cases of uncertainty to promote discussion;*
- *beware imprecise usage, such as 'catching a flight'.*

Of the three terms that we are using here, defining an entity type is probably the most difficult. Having said that, in practical terms maybe 70 per cent of the entity types of importance will be self-evident – CUSTOMER, ACCOUNT, ORDER etc – but the remaining entity types might take considerable effort to discover. In general terms then, an entity type is a collection of attributes that, when taken together, has meaning in its own right. A handy rule-of-thumb here is to look for nouns in the description of a business area or function. If you are interviewing users and ask them to describe what they do, the nouns used in the description will basically correspond to entity types, for example:

> "We receive ORDERS from CUSTOMERS. Each ORDER consists of a number of ORDER LINES each referring to a different PRODUCT. The CUSTOMER must have an ACCOUNT with us and this will be debited with the INVOICE for the goods."

This may not give a totally valid list of entity types, but it is a good starting point for the data analyst, especially if knowledge of what happens in that part of the business is sketchy. Applying the techniques outlined in Chapters 3-5 to this starting point will help to establish if other entity types exist, and where they fit into the structure.

Most entity types can be seen to exist in the outside world and are fairly concrete, such as CUSTOMER, PART. In other words an entity is something you can kick. But equally valid are many abstract entity types which are equally obvious: ACCOUNT, DEPOSIT, WITHDRAWAL. The rule comes down to whether or not the concept is felt to be 'real' to the organisation and there are obviously no hard and fast rules on what should be an entity type. The data analyst should not be afraid of creating entity types where it is felt that this might be useful. It is far easier to combine entity types together if required than to split them up. The technique also promotes discussion and is more likely to highlight the data requirements.

Again, it should be remembered that we are attempting to establish a data structure that is independent of the eventual implementation, so the complexity of the structure should not be an inhibiting factor if that complexity is justified to represent the data.

Finally, one should bear in mind the many definitional problems that can arise when trying to work out what the entity type is. In everyday parlance, people talk quite naturally, for example, about catching a plane or a flight but neither is really correct.

A basic data model to show this situation is given in Figure 2.4. The FLIGHT, as known by the flight number, occurs daily or weekly or whatever, and the data about a FLIGHT is the same regardless of the date on which it happened. The PLANE on the other hand will have certain fixed detail associated with it, but it may service any FLIGHT on any particular DATE. Now it is also apparent that there will be some information that is associated purely with a FLIGHT by a particular PLANE on a particular DATE and this detail also needs to be recorded. This fairly simple example serves to illustrate the sort of situation that can be misleading if the data analyst does not probe deeply into the underlying meaning of the terms being used.

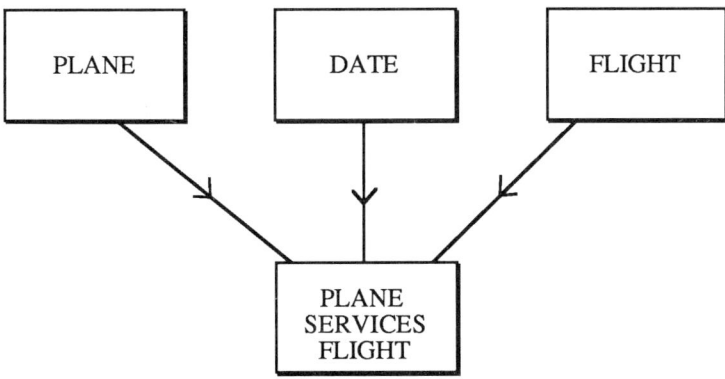

PLANE [Plane Type, Plane Number, Number of Seats etc]

DATE [Calendar Date]

FLIGHT [Flight Number, Departure Time, Departure Airport etc]

PLANE SERVICES FLIGHT [Plane Number, Flight Number, Date]

Figure 2.4   The relationship between PLANE, FLIGHT and an actual flight on a particular date

Attributes

Attribute may take a number of distinct roles, as follows:

- *Descriptive: these attributes add meaning to the entity type; most attributes are descriptive;*
- *Identifying: one or more attributes in an entity type must serve to identify uniquely each entity occurrence, these attributes may also be descriptive;*
- *Supportive: these attributes support relationships between entity types; they may also be descriptive or identifying attributes.*

Let us begin by restating the definition of an attribute. Attributes are those individual bits of data that go to make up an entity type. This is a rather loose definition, but it will serve our purposes. As a general rule, an attribute has little meaning unless it is combined with other attributes in an entity type; an address, for example, is fairly meaningless without knowing to which CUSTOMER it refers.

A problem for many inexperienced data analysts comes in deciding whether something should be an entity type or an attribute. This is a perennial problem and there is no glib answer. Applying the techniques outlined in the following chapters will help to determine the answer, but there will always be situations where interpretation has to be made and a conscious decision taken. It is important though to note where this has been done, so that the area can be reviewed to establish if further clues have emerged as to the correct answer.

The same attribute may occur in several entity types and there is nothing wrong in this. Much more worrying is where the same attribute appears several times in the same entity type. This is usually, but not always, an indication that something is amiss and that further analysis needs to be done.

In some approaches to data analysis the concept of 'domain' is used. This is normally used where the data analyst wishes to indicate that certain attributes occurring in different entity types are of the same nature but that, depending on the entity type in which it is used, the permissible values may be different. Let us take the example of 'height'. This attribute could occur in the entity type PERSON, or MOUNTAIN, or BUILDING, but the allowable values of the attribute will differ depending on the entity type in which it occurs. All values of height will be drawn from the same domain of values, that is, some measure of length. Thus values such as 25,000 feet, 3 miles, 5 feet, etc, would be valid for the Domain, whereas 2 gallons would not. The allowable values for the attribute, however, will be a subset dependent on the entity type in which it occurs. 25,000 feet is valid for MOUNTAIN, but not for PERSON.

This concept is not always of great use in day-to-day data analysis but can be useful where some mechanised process of data model documentation is being used.

However, attributes do more than just combine to make up an entity type. They have three distinct roles which they may fulfil and in some cases, an attribute may function in more than one role. The first of these roles we have already looked at in passing and that is the Descriptive Role. This quite simply means that the attribute says something about the entity type in which it occurs, that is, it adds meaning. Most attributes fulfil a Descriptive Role and the value taken by the attribute in a particular occurrence of the entity type may be non-unique.

Secondly, an attribute may function in an Identifying Role either by itself or in combination with other attributes. It is a cardinal rule of data analysis that each occurrence of an entity type must be uniquely identified and distinguishable from all other occurrences of that entity type. For a given entity type it might be possible to have one attribute which will uniquely identify each occurrence but, in many cases, two or more attributes may need to be combined to achieve this uniqueness. Most Identifying attributes are also Descriptive.

Lastly, an attribute may function in a Supportive Role. That is, the attribute supports a relationship from one entity type to another. This concept will become clearer in the discussion on relationships in the next section, but for now it can be assumed that, where a relationship exists between two entity types, there must be one or more attributes that are common between the two and these attributes must also function as Identifiers in at least one of the entity types. This is a very important requirement and it serves to identify the occurrences of one entity type that are subordinate or 'belong to' an occurrence of another entity type. Supportive attributes may also function as Descriptive attributes and sometimes as Identifying attributes.

As that was rather academic and theoretical, let us try and put it into real terms. A portion of a data model concerning Cars and Car Makes is shown in Figure 2.5, along with the appropriate attributes. The figure also shows the notation that can be used to indicate different roles each attribute fulfils.

The MAKE entity type is identified by an attribute called Make Code which is deemed to be unique for any occurrence of MAKE. All other attributes are purely descriptive.

The MODEL entity type is identified by two attributes, Make Code and Model Code. It is assumed here that the Model Code on its own would not be unique and that the Make Code is required to provide this uniqueness. The Make Code also functions in a Supportive Role. It indicates which occurrence of the MAKE entity type a particular occurrence of MODEL relates to. All other attributes are purely Descriptive.

The CAR entity type is somewhat different. The entity type is identified by an attribute called Registration Number which will identify a particular vehicle. In the real world, of course, Registration Numbers can change, but for the purposes of our example let us assume that this cannot happen and the Registration Number is unique and sufficient to identify an occurrence of CAR. The relationship to MODEL has also to be supported and this is achieved by propagating the identifiers of the MODEL entity type into the CAR entity type. In this case, the Supportive attributes do not function as Identifiers. All attributes are Descriptive.

This should have served to illustrate the roles that an attribute can fulfil, and further details concerning which attributes are Identifying or Supportive and how to use them will be apparent in later chapters.

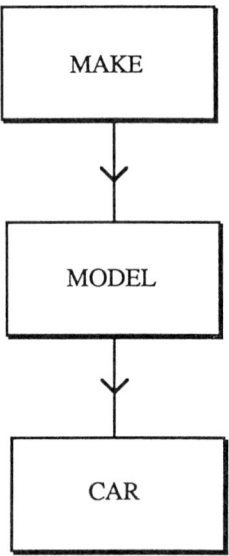

MAKE   [Make Code, Make Name, Country of Manufacture, etc]

MODEL   [Make Code, Model Code, Model Description, Number of Seats, Type of Body, Etc]

CAR   [Registration Number, Make Code, Model Code, Year of Manufacture, etc]

——— = Identifying attributes
- - - - = Supportive attributes

Figure 2.5   Attribute Roles; Identifying, Supportive and Descriptive

Relationships

Each type of relationship is significant:

- *One-to-One : each occurrence of one entity type may be related to only one occurrence of the other entity type;*
- *One-to-Many: each occurrence of the entity type at the 'many' end of the relationship must be related to one, and only one, occurrence of the other entity type;*
- *Zero-or-One-to-Many: the occurrences at the 'many' end may be related to an occurrence of the other entity type but do not have to be;*
- *Many-to-Many: this is a useful initial relationship to use but should be broken down as analysis proceeds;*
- *Recursive: a special type of relationship, taking any of the above forms when occurrences of the entity type are related to other occurrences of the same entity type.*

Perhaps the most significant concept within data analysis is that of relationships between different entity types. A relationship can be defined as a logical connection or dependency between occurrences of one entity type and occurrences of another. There are, however, a number of different types of relationship, each of which merits discussion on its own account.

One-to-One Relationships

The simplest form of relationship is shown in Figure 2.6 and is known as a One-to-One Relationship. What this signifies is that for an occurrence of one entity type there may be an occurrence of the other entity type. But more importantly, for an occurrence of one entity type there *may not be more than one* related occurrence of the other entity type. In our example we have taken an entity type containing basic account details and a second entity type which can hold additional optional data. It is apparent that there can only ever be one chunk of additional detail for an account, and furthermore, that for a chunk of additional detail there can only be one account.

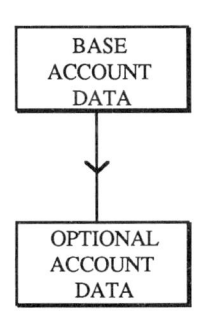

Figure 2.6 A One-to-One Relationship

Data Analysis Concepts

In practice however, One-to-One Relationships rarely occur in the real world. There is little advantage in keeping the two entity types separate. They would usually be combined into one entity type with the attributes from both.

One-to-Many Relationships

Much more common and far more useful is the One-to-Many Relationship as shown in Figure 2.7. The relationship signifies a number of points. In this example, it states that there is a relationship between the CUSTOMER and ACCOUNT entity types. But it also says much more. The arrow points from 'the one' to 'the many'. For each occurrence of the CUSTOMER entity type there may be many occurrences of ACCOUNT. The meaning of the phrase 'many' must be clarified. In this context, 'many' signifies zero, one or more. Obviously, a CUSTOMER may not have any related ACCOUNTS, a CUSTOMER may have one related ACCOUNT, or alternatively, a CUSTOMER may have several ACCOUNTS. For each CUSTOMER occurrence, one of these three conditions must always be true. But perhaps the most important information is gleaned from following the line from the 'many' to the 'one'. Here the relationship means something quite specific. It states that for any one occurrence of the entity type ACCOUNT there *must* be *one, and only one*, related occurrence of the CUSTOMER entity type. With our simple example, this is fairly self-evident. Although a CUSTOMER can have any number of ACCOUNTS, an ACCOUNT can not exist unless there is a related CUSTOMER, and also, an ACCOUNT cannot be related to more than one CUSTOMER. The constraints implied by this type of relationship are very important and cannot be overstressed.

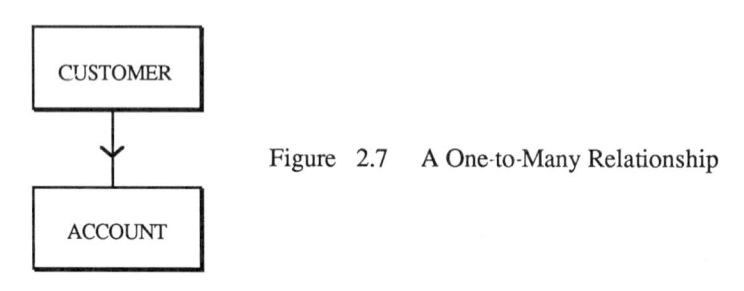

Figure 2.7   A One-to-Many Relationship

Zero-or-One-to-Many Relationships

The next type of relationship we should consider is known as a Zero-or-One-to-Many Relationship and is shown in Figure 2.8. The relationship has a similar meaning to the One-to-Many Relationship when interpreted in the direction of the arrow, that is, from the 'one' to the 'many'. A SCHOOL *may* have zero, one or more CHILDREN, although in point of fact a SCHOOL without any pupils might be a little unlikely! However, the meaning of the relationship is very different when viewed from the 'many' to the 'one'. The dotted line signifies that for any occurrence of CHILD there may be, but does not have to be, a related occurrence of SCHOOL. But, if there is a related SCHOOL, there may only be one. The significant point here is that both entity types, although related, are independent of each other.

Figure 2.8   A Zero-or-One-to-Many-Relationship

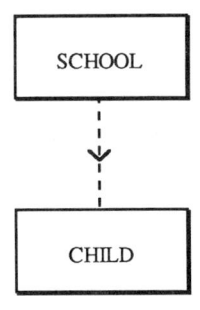

In the case of One-to-Many Relationships the 'many' is entirely dependent on the 'one'. An ACCOUNT, to take the previous example, *must* have an associated CUSTOMER if it is to exist at all. In the Zero-or-One-to-Many Relationship, both entity types can exist independently but may be related. An occurrence of the entity type CHILD can exist regardless of whether or not there is an associated SCHOOL. The Zero-or-One-to-Many Relationship is relatively common although not as frequent as the One-to-Many Relationship. The number of these found in a data model tends to decrease as analysis continues and the analysts understanding of the problem areas deepens. It is necessary to examine carefully each Zero-or-One-to-Many Relationship as they tend to indicate that the analysis is not complete and that other facts need to be uncovered.

Many-to-Many Relationships

The next type of relationship to be examined is known as a Many-to-Many Relationship and is shown in Figure 2.9. This is again fairly straightforward to understand, stating that any occurrence of the POLICY entity type may be related to zero, one or more VEHICLE entity occurrences. The relationship reads both ways, though, as each occurrence of VEHICLE may be related to zero, one or more occurrences of POLICY. Many-to-Many Relationships are very common. When an area is under initial investigation, Many-to-Many Relationships tend to appear between the majority of the major entity types. Although Many-to-Many Relationships are a very convenient and easy way of thinking about related entity types, in practical terms, a Many-to-Many relationship actually tells us very little about the particular situation that is

Data Analysis Concepts

being modelled. There is also considerable debate about whether Many-to-Many Relationships actually exist in the real world or whether they really indicate something else. In our example, the relationship tells us nothing about *when* each VEHICLE was covered by a particular POLICY or vice versa. In order to make the whole thing more meaningful, it is necessary to carry out further analysis to discover what actually underlies the situation. The results of this are shown in Figure 2.10. Here we have introduced a third entity type, VEHICLE ON COVER, and broken the Many-to-Many Relationship down into two One-to-Many Relationships. The new entity type will have an occurrence for each period that a given VEHICLE is covered by a particular POLICY. This has, in effect, created an entity type to record details about each occurrence of the Many-to-Many Relationship itself. This is obviously much more meaningful and useful.

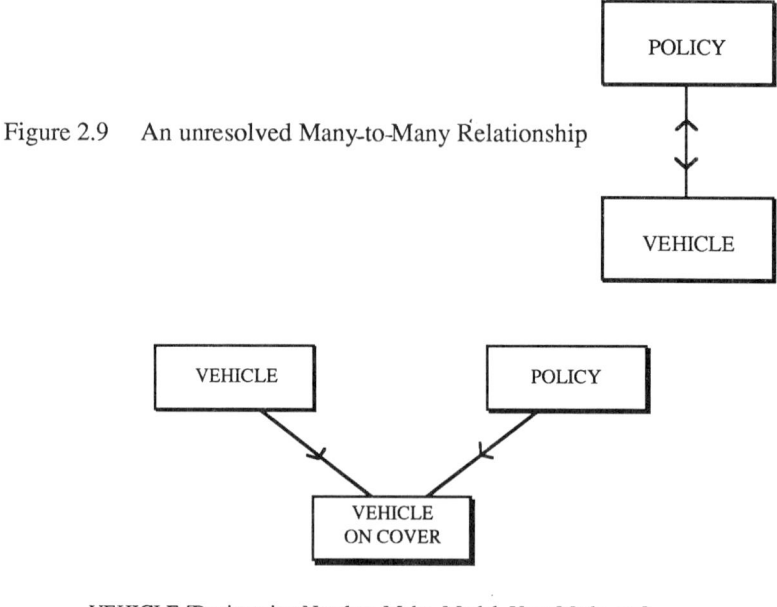

Figure 2.9   An unresolved Many-to-Many Relationship

VEHICLE [Registration Number, Make, Model, Year Made, etc]

POLICY [Policy Number, Inception Date, Type of Cover, etc]

VEHICLE ON COVER [Registration Number, Policy Number, Start Date, End Date]

Figure 2.10   A resolved Many-to-Many Relationship with basic attributes

It is also true that an occurrence of VEHICLE ON COVER can only exist if it relates to a particular POLICY and a particular VEHICLE. We have assumed for convenience that a VEHICLE will only start to be covered by one POLICY on a given date, and that therefore only these attributes are required as Identifiers for the VEHICLE ON COVER entity type. It is now possible to trace which VEHICLES were covered by a POLICY and when, and for a VEHICLE which POLICIES have covered it and when. For this relatively concrete example, the resolution of the Many-to-Many Relationship is fairly obvious, but in some cases it can take much more thought and analysis to discover exactly what structure underlies the apparent Many-to-Many.

The foregoing then, implies that Many-to-Many Relationships are of little use in practical data analysis, but this is probably an over statement. Certainly, in the early stages of analysis when the data analyst is still thinking in fairly broad terms about the problem area, Many-to-Many Relationships can be useful stepping stones in conceptualising the basic structures. However it must be borne in mind that as the level of detail increases these relationships will need to be re-examined and resolved. Many-to-Many Relationships are also of use where the data analyst wishes to prepare an overview of the results of the analysis and does not wish the data model to become over complicated with what may, at first sight, appear to be fairly esoteric or contrived entity types. They should be used with care, though, as the simplification that they allow can hide problems and prevent these being addressed.

Recursive Relationships

The final type of relationship to be discussed is known as a Recursive Relationship and is shown in Figure 2.11. This type of relationship allows occurrences of an entity type to be related to other occurrences of the same entity type. This type of structure appears quite frequently, typically in Bill-of-Material applications, organisation structures and other hierarchial systems. Taking Bill-of-Materials as our example, we are really saying that any particular PART may be made up of a number of other PARTS, and that each PART may be used to construct any number of other PARTS. The relationship itself may be of any of the four preceeding types, One-to-One, One-to-Many, Zero-or-One-to-Many, or Many-to-Many. The line representing the relationship will, of course, take the appropriate format for that type of relationship.

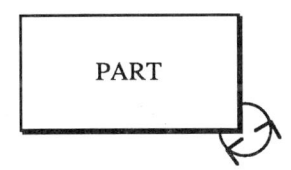

Figure 2.11   Representation of Recursive Structure

As with Many-to-Many Relationships, Recursive Relationships are a useful short-hand for overview and broad-level views but they have their own particular difficulties and need to be broken down. In the case of Recursive Relationships, the considerations in resolving the structure are more involved than those for Many-to-Many Relationships and a full discussion of the technique is to be found in the following chapters in the sections dealing with hierarchies, organisation structures and other recursive entity types.

General Points

Three general points concerning relationships are worthy of note. Firstly the function of the arrow-head on the line does not imply any form of flow from one entity type to another. The sole purpose of the arrow is to distinguish the 'many' from the 'one'. The arrow always points towards the many and away from the one. Some data modelling conventions use other symbols in order to avoid this potential confusion, but throughout this book arrows will be used as described.

Secondly, there is no necessary concept of ownership of the one over the many. This is a common misunderstanding by people unfamiliar with the technique, and it is quite understandable. In many cases a relationship does imply ownership but this is only in the context of the entity types that it links. Obviously, an ACCOUNT cannot exist without a CUSTOMER (Figure 2.7) but it is not really correct to say that the ACCOUNT belongs to the CUSTOMER. Again, it is quite useful to consider most relationships as linking a 'superior' entity type to occurrences of a 'subsidiary' one, but the data analyst must remember that this may not always be the case.

To complete this section on relationships, it is necessary to discuss some of the points concerning relationships that are not made explicit by the modelling convention that we have adopted. Although these points are fairly minor they should be borne in mind by the data analyst. The first of these concerns 'essential members'. In discussing One-to-Many Relationships, the point was made that the relationship implied that an occurrence of the 'many' could not exist without a related occurrence of the 'one'.

The converse is not true. The relationship does not imply that the 'many' must exist for each occurrence of the 'one'. In other words, the 'many' are optional. Unfortunately, this is not always true. There are situations where the 'many' must exist if the 'one' is to have any meaning. The modelling convention, however, does not allow us to distinguish between the two situations. Where it is necessary to make the distinction clear, this should be made in the back-up documentation for the model which will be discussed in later chapters.

The second point concerns 'exclusive relationships'. Situations will often arise where an entity type participates either as the 'one' or 'the many' in more than one Zero-or-One-to-Many Relationship and this is quite permissible. However, in some cases, participation in some or all of these relationships is interdependent. Typically this is where each occurrence of an entity type may, or has to, participate in any one of these relationships, but can only participate in one. In other words, participation in these relationships is exclusive. Again, this situation cannot be shown explicitly in the modelling convention we have adopted. Once more, where this needs to be explicitly mentioned the point must be made elsewhere in the back-up documentation.

We have now covered the basic underlying concepts of data analysis. It is important that these concepts are clearly understood in order to interpret the techniques in the following chapters.

Chapter 3

Basic Data Analysis Techniques

Concentrate on Business Data

Data analysis concentrates on the business view of data. Any current or planned systems should not influence the analysis.

Although data analysis is to an extent independent of process analysis, it is necessary to use processes to scope the data analysis exercise and to define boundaries.

The data analysis techniques contained in the next three chapters should be applied with the general aim of arriving at a data model that describes the data of interest to the organisation at the business level. It is very important that the data analyst is divorced from the influence of existing or potential systems. *Data analysis is independent of any implementation and should concentrate solely on the business view of its data.* Not only is the shape of the implemented solution irrelevant to the data modelling exercise, but also any hardware or software considerations should be ignored at this stage.

In this way, the data analyst is attempting to develop a much purer statement of what is required by the organisation than would be the case if systems design had proceeded on conventional lines. It is possible to make a theoretical argument supporting the view that data analysis can be carried out without *any* regard to the processes that will use that data. This view is not supported here. Some boundaries have to be drawn to set the scope of the exercise and data must be useful and pertinent or there is no point in considering it. The interplay between data analysis and process analysis is considered later in the text, but for now it is sufficient to consider processes merely as a means of preventing the data analyst going off on a tangent and enabling us to record some indication of whether a piece of data *may* or *may not* be of interest.

The following techniques can be roughly grouped into three categories.

Firstly: techniques to use when producing the data model, described in this chapter.

Secondly: useful approaches to refining and rationalising a data model to eliminate redundancy and over complexity. These can be found in Chapter 4.

Finally, techniques to ensure that attributes are correctly assigned to the right entity types and that the data model hangs together, are considered in Chapter 5.

Time Dependency

Although a snapshot view may be a useful initial technique, the effect of time must be accommodated. Typically, this will:

- *complicate existing relationships;*
- *cause the creation of additional 'time dependent' entity types;*
- *add additional attributes as date stamps.*

Once the effects of time have been considered then the model becomes more robust and flexible:

- *the timescale becomes infinite and independent of storage capabilities;*
- *historical enquiries and time analyses are made possible;*
- *by considering time issues, entity types may be uncovered which will prove useful in other ways and lead to a more complete model.*

One of the most important aspects of data analysis which must never be far from the mind of the data analyst, is the effect of time on the data structures that we are trying to establish. It is extremely important to avoid modelling a situation which is merely a snapshot view of the data. It is often quite useful to establish the major entity types and basic relationships by actually making use of exactly this type of viewpoint, but the data analyst must be fully aware of the potential dangers of this. A simplistic, snapshot view is often characteristic of conventional systems where data can be overwritten or discarded as and when the situation changes. This makes things very difficult as soon as users require information about anything other than the current situation. With data modelling, the effect of time on the data is even more fundamental. Changes over time can not only affect what entity types are needed and therefore their attributes, but also the relationships and the required data structure. These effects will be examined in a little while, but first, we should discuss some of the questions that the data analyst needs to bear in mind.

First, consider the future. If this is the situation now, will it always be so? If not, is it important to record the changes? Assuming the answer to the first question is no, that is the situation will change, then it is necessary to ask the second. Broadly speaking, if it can be established that the new situation is a simple change (perhaps as a result of a prior error), and it is not necessary to date-stamp the alteration, then the effect on the data model may be zero save that of allowing the change to occur.

However, when these situations are carefully examined it is usually necessary to ensure that the change can occur, will be recorded as having

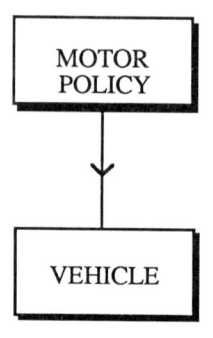

Figure 3.1   An 'apparent' One-to-many Relationship

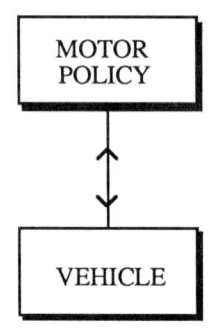

Figure 3.2   A 'Many-to-Many' Relationship created by the effect of time

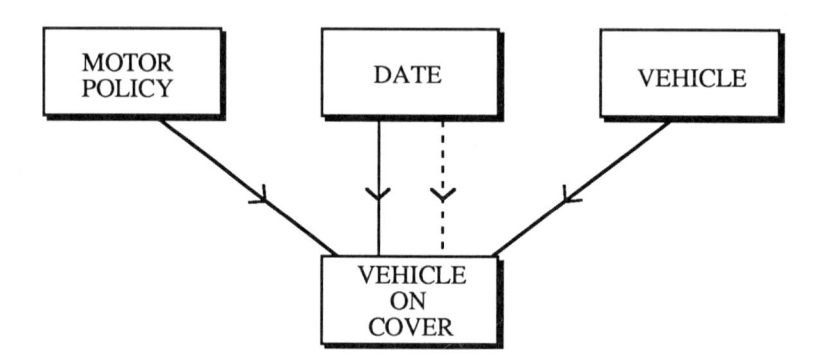

Figure 3.3   Resolution of the Time Dependent Many-to-Many

happened and that it will be possible to establish the 'before and after' conditions. It should also be remembered here that there is often a requirement to record some future event that will occur, such as a request to take some action in a month's time. In this situation there are two distinct requirements: firstly, to record what the action is to be and when it is to be done; secondly, the need to take the appropriate action when the future date arrives and, normally, record that it has been done. It is common, therefore for DATE to be a useful entity type and there are normally many relationships between DATE and other entity types.

In a similar vein, it is also necessary to consider reflecting things that have happened in the past. Once more it is necessary to examine entity types and relationships and ask if the situation will change and if the history of change is important. History can be very useful and once more it is usually necessary to record change. Therefore it is unsafe to take an oversimplistic view of time and data.

In practical terms, how does this affect relationships? First, consider the situation shown in Figure 3.1. The structure implies that many VEHICLES can be covered by a MOTOR POLICY, and additionally for each VEHICLE there will be one MOTOR POLICY. This is fine for a snapshot view of the situation, but when we start to consider the effect of time on this relationship it plainly does not satisfy the requirements. Over time a given VEHICLE may be covered by many MOTOR POLICIES, and the details of these changes would be very important and would have to be recorded.

At first sight then, there is a Many-to-Many Relationship as shown in Figure 3.2 which will need to be resolved. The resolution is depicted in Figure 3.3. This includes not only an additional entity type of VEHICLE ON COVER but also two relationships from that entity type to DATE, one of these relationships being optional, a Zero-or-One-to-Many Relationship. So what does this model tell us? Firstly, it can be seen that a given MOTOR POLICY may have a number of VEHICLE ON COVER entity occurrences each of which will indicate a VEHICLE that has been, is, or will be covered by that MOTOR POLICY. Starting at VEHICLE we can also discover which MOTOR POLICIES have covered, will cover, or are covering a given VEHICLE. The importance comes in considering the VEHICLE ON COVER entity type. Each occurrence of this entity type will identify one MOTOR POLICY and one VEHICLE and this combination will have been in force, is in force or will be in force in the future. These last few options are determined by the relationship to DATE. Why two relationships? Obviously, there must always be a 'start date' for the combination whether it is past, current or future. However, logic dictates that there would not always be an 'end date' for when a VEHICLE would no longer be covered by a MOTOR POLICY. The VEHICLE ON COVER entity type will have both attributes, one supporting each relationship, and Start Date will function as an Identifying attribute, along with, say, Policy Number and Registration Number. The optional nature of the second relationship simply means that the End Date attribute will not have a value in all occurrences of the entity type.

There are many occasions in data analysis where one encounters situations similar to the above. The effect of future and past events on relationships is worthy of separate treatment and is discussed in a later section of this chapter.

It is generally important for the data analyst to watch out for 'date-stamps' when determining the attributes of an entity type. Any such date-stamp implies that a time dependency exists and that the entity type and/or its relationships can change over time.

Data analysts often become concerned over the timescale that should be considered when carrying out data analysis. This is really a red herring as once the effect of time has been allowed for in the data structure, the structure will support an infinite timescale without change. How long data remains relevant to an organisation is process dependent and does not impact the modelling exercise in any way. Obviously, in the long run of actually designing a system and the data and processes that go with it, this question will have to be answered. The techniques for establishing this and the questions that need to be answered are discussed much later in this book. Similarly, questions often arise concerning the archiving of historical information and queries on how long data should remain on-line. This is pure implementation and again does not affect the data model.

To sum up this section, then, two general guidelines become clear. First, all entity types and their relationships must be examined to establish the effect of time upon them: apparently simple structures can become, quite correctly, more complex once this factor is considered. Secondly, the data model should reflect an infinite timescale and all questions of how long data need be held become process dependent rather than time dependent.

Categorisation of Entities

Most categorising entity types are obvious. Clues to their existence are:

- *a large number of entity occurrences for the entity type at the top of the structure;*
- *many entity occurrences showing the same value for a specific attribute.*

There may be interrelationships between categorising entity types and these can affect relationships at a lower level. The use of categories may enable relationships to be expressed at a higher level in the structure which will typically tighten constraints on data entering the system.

The following techniques concern themselves with an area of data modelling that is all too often ignored, not just by conventional systems, but by many data analysts. By examining the data, it is possible to identify many 'implied' entity types that identify subsets of the main entity types. If these are recognised explicity they serve to categorise the larger entity types and allow easier, more precise access to the data structure. A categorising entity type, is any entity type that serves to define subsets of occurrences of another entity type. All too often systems are built around an amorphous mass of data records with little or no explicit structure save the odd data type or whatever. Again, this is partly the effect of the limitations of conventional file handling techniques but we now have the opportunity to consider this aspect of data organisation more fully.

But why should we bother? It is a preoccupation of mankind to try and bring some order to the universe. People do think in terms of categories and classifications. In this sense, the categories themselves become real and are therefore valid for data modelling. There are frequently requirements to present data in some structured way and these categorisations should be held explicitly wherever possible. The major benefits are easier enquiries and well-structured updates with less chance of error. Many of the rules that stipulate what data values are acceptable can be reflected at the category level and then become much more meaningful. The implications of this will be considered in the next chapter, but first let us consider types of categories more fully.

There are, of course, many obvious categories that can be defined very easily. Most of these will be expressed quite naturally and in some of the prior examples we have already met some instances of this type of category. In Figure 2.5, there were two categories classifying VEHICLES, namely MODEL and MAKE. These straightforward categories are typical of these

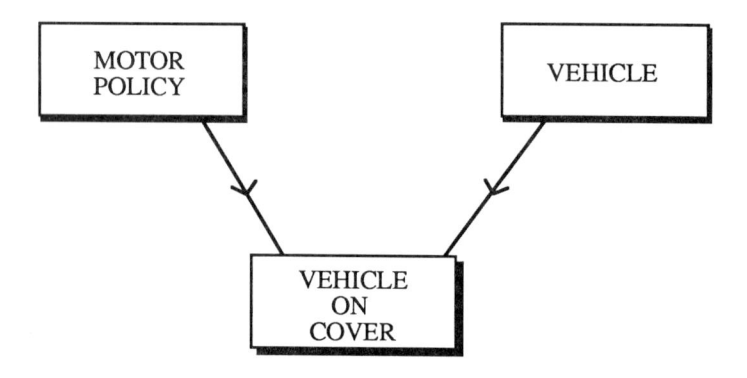

Figure 3.4 Simple structure with no Categorisation

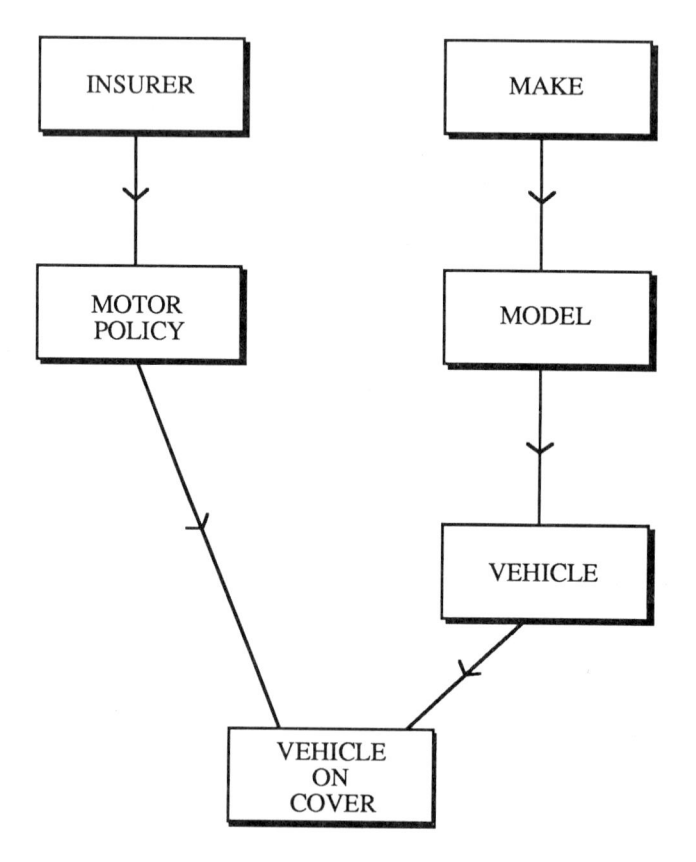

Figure 3.5 An enhanced structure with Categorisation

obvious categorisations. It is reasonable to suppose that any organisation that needs to record details of VEHICLES would probably want to classify those details for reports or statistics by this sort of category. But there is more benefit to be drawn from this type of categorisation. The nature of the relationships implies that to store details of a VEHICLE, the MAKE and MODEL must already exist. It would not be possible, for example, to give a VEHICLE a MAKE and MODEL of, say 'Ford' and 'Mini' unless this was already set up. If the categories are not made explicit then this and other incorrect values could be entered. We can also save on data storage. If MAKE and MODEL are not created as separate entity types then each VEHICLE will have to hold this data and the model description would be stored as many times as there were VEHICLES of that type.

When analysing an area there will be many of these obvious categorisations, but there will also be many that are less obvious or implicit. The data analyst will normally have to work a little harder to uncover these implied categories and, having made them explicit, may have difficulty in persuading users that they do exist. So how do we set about uncovering these categories? There are two main clues to their existence.

Firstly, consider the volume of entity ocurrences at the top of the data structure, there is a familiar example in Figure 3.4. This is obviously an oversimplistic model but it illustrates the point quite nicely. Assume that this is part of a data model for a Motor Insurance system, and now consider how many occurrences of both MOTOR POLICY and VEHICLE will exist. It is obviously quite a large number in both cases. As has been said, people try to categorise the world and will order large numbers of things into fewer types.

Figure 3.5 shows the same model with a number of additional entity types. INSURER has been added to extract this information, and MAKE and MODEL have been added above VEHICLE. In this example, the categories are fairly obvious but there are also many other ways that both VEHICLES and MOTOR POLICIES may be categorised. However, even if this had not been so obvious, the clue to their existence would have been the *number* of VEHICLES and the *number* of MOTOR POLICIES. Both would have been of sufficient magnitude to *suggest* that some form of category had been omitted.

The second clue to the existence of categories is the repetition in many occurrences of an entity type of the same *value* for a particular attribute. If the value of an attribute is often repeated and if, more importantly, there are only a finite number of possible values for the attribute, then this is likely to be something which should be taken out into a separate entity type and used to categorise the original entity type. In the previous example MOTOR POLICY was a good illustration of this. Insurer code, or a similar identifier, would have certainly been an attribute of MOTOR POLICY in Figure 3.4.

But the range of values that are valid is limited and the same values would have occurred many times in the hundreds or thousands of occurrences of MOTOR POLICY.

Categories also have justification in terms of listing data and, similarly, searching for it. If data is to be listed or summarised by the system then almost certainly categories will be needed in order to group like with like. If this requirement exists, the categories will almost certainly be easy to obtain and build into the data model. In other situations a little more digging will be necessary to establish what they are: search requirements are a little more complex. To stick with the previous example, if there was a requirement to search for a particular MOTOR POLICY then several thousand occurrences may have to be examined before the correct one is found. By including the INSURER explicitly there is now a mechanism for reducing the length of the search and thus make the whole requirement easier to manage. Where search requirements are more complex the data analyst may have to go to some lengths to determine what categories exist and how they should be related in order to structure the data to best advantage.

This brings us to the point concerning the number of levels of categorisation that are needed. Where the number of entity occurrences is large, several categorisations may be required, each categorising the one beneath. This structure can be built-up progressively by asking the same questions about each category as were asked about the lower level entity type itself. But when building these structures another vital question needs to be asked. It is quite possible to 'miss' a level of categorisation and find that, although there are only a few occurrences of the higher entity type, there are still too many occurrences at the lower level for each occurrence at the higher level. Having discovered a higher level categorising entity type it is necessary to look at the relationship between the two to ensure that an intervening entity type has not been missed.

It is also quite possible for an entity type to be categorised in several different ways and for these categorisations to be interrelated. Care must be taken here that the right levels of classification have been used and to guard against the same categorisation appearing under different names. These points are more fully dealt with in Chapter 4.

The data analyst should be prepared to 'invent' categorising entity types where this appears useful as this will, at least, provoke discussion of the area and highlight the problem. Again, it is far better to discuss the entity type and then discard it than not to be aware of it at all. There are also some catch-all phrases that can be used here for naming these entity types. Words such as Category, Class and Type are very useful but if a more detailed breakdown is required these can be further qualified by 'Broad' or 'Precise'. It is, of course, very important to ensure that any 'invented' entity types are fully discussed and not just inserted into the model on the off-chance.

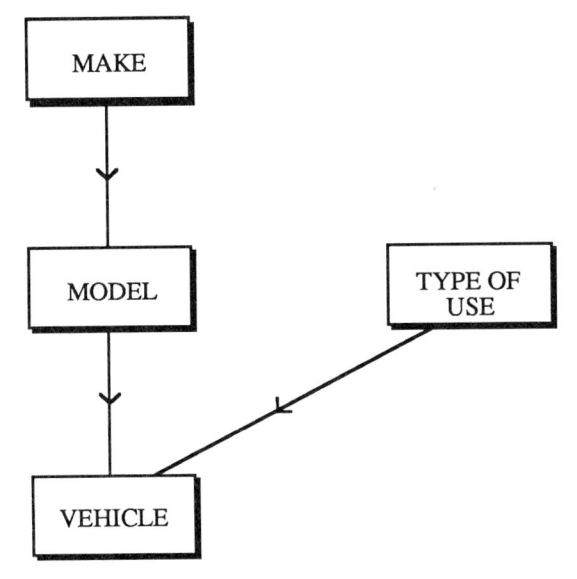

Figure 3.6 Both Categorisations at the Base Level

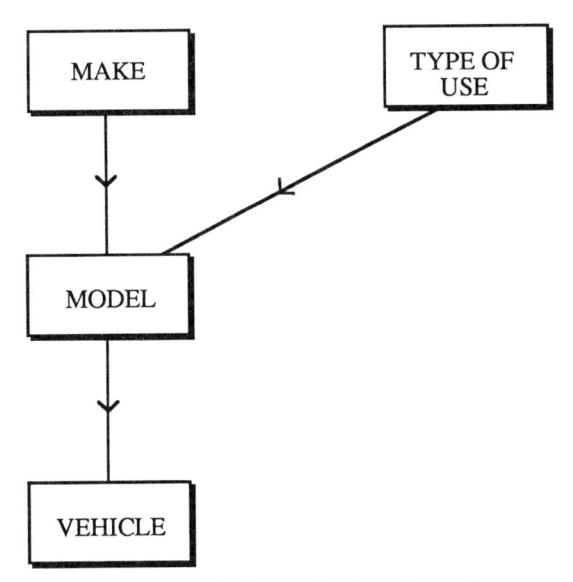

Figure 3.7 Both Categorisations Interrelate

Basic Data Analysis Techniques

The last point concerning categorisation is the effect that they have on the relationships in which the entity type participates. If we re-consider the earlier example of VEHICLE, MAKE and MODEL and assume the need to indicate the type of use – Passenger, Light Goods, Heavy Goods or whatever – then this can be shown as in Figure 3.6. In this structure VEHICLE is categorised in two ways, and every time a new occurrence of VEHICLE is created it is necessary to specify the MAKE, MODEL and TYPE OF USE. The effect of this would be to allow a 'Mini' for example to be described as Heavy Goods which is clearly wrong. By moving the relationship between TYPE OF USE and VEHICLE, to TYPE OF USE and MODEL, as shown in Figure 3.7, this can be resolved.

In other words the categorisation system itself raises the level at which the 'rule' can be expressed. Now when creating VEHICLE, the specification of the MODEL will predetermine the TYPE OF USE that applies. Further discussion of this topic can be found in Chapter 4.

Events

Events have two major implications:

- *the event may change data items or their relationships; the changes will need to be recorded;*
- *the event itself may need to be recorded and will therefore have associated data; this data will need to be captured in an appropriate entity type.*

So far this book has concentrated on the data that needs to be held and is of interest to an organisation. An earlier section in this chapter discussed time-dependency – the degree to which data either changes or is dependent upon the passing of time. But this is only half the story. *Things happen in any business that fundamentally affect the data itself. These 'events' are of interest to the data analyst and need to be recorded.* So, for our current purposes, an event is something that causes data to change and furthermore, an event usually occurs at a specific point in time.

From the data modelling point of view, these events need to be modelled and included in the structure usually for one of two specific reasons. Firstly, data has been affected by an event and it is necessary to record the change and also the fact that it happened. Secondly, the happening of the event itself has value and we need to record that the event has occurred.

The first of these is really an extension of the time-dependency argument, and the Motor Insurance example can be extended to illustrate this. Each policy will have a different status at different times: New Policy, Adjustment Outstanding, Renewal Invited, Lapsed, etc. Each change of status is an event that happens at a particular time and needs to be recorded for obvious reasons. The most straightforward way of modelling this is shown in Figure 3.8. This model will quite easily record what status a particular MOTOR POLICY occurrence has, but little else. It does not record when things happened or the sequence and history of those events.

In order to model this effectively the structure shown in Figure 3.9 is needed. This allows us to say not only when each event occurred for a particular occurrence of MOTOR POLICY, but also the history of these events. It is also possible to record future events with this type of structure, or it could be used to enable the system to trigger event-dependent changes on a particular date. This provides a means of extracting and recording data relevant to the particular happening of an event to a given occurrence of an entity type and it is always possible to trace what happened and when. Figure 3.9 only shows one relationship between DATE and POLICY HISTORY. This is adequate providing MOTOR POLICY entities only ever have one POLICY STATUS at a point in time and that they always have

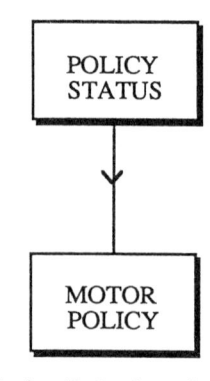

Figure 3.8 A simplistic view of event modelling

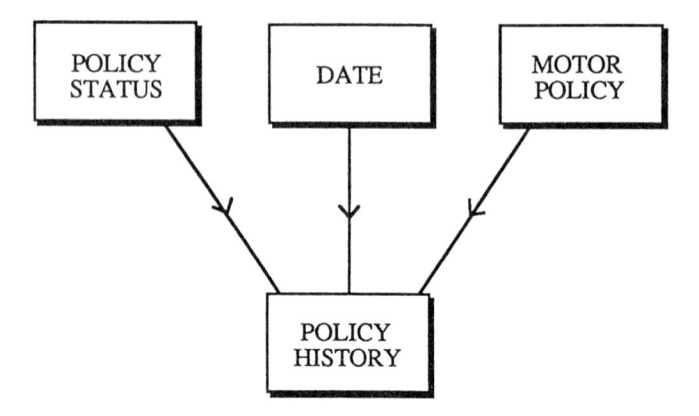

Figure 3.9 An enhanced 'event recording' structure

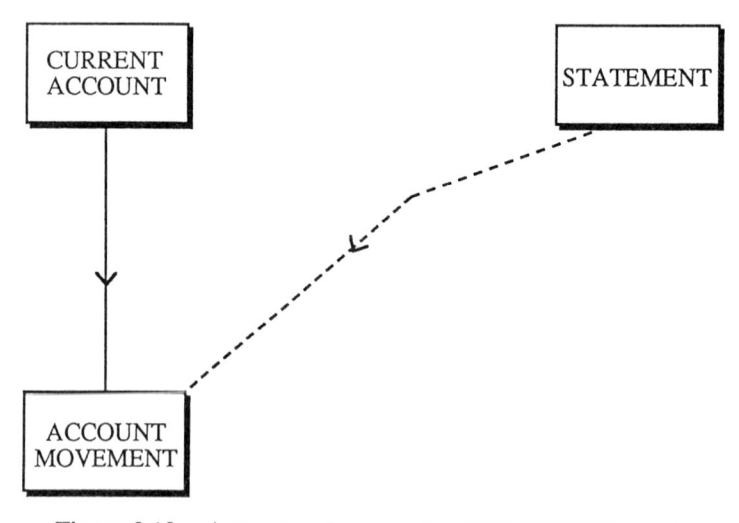

Figure 3.10 A structure for recording 'STATEMENT' events

one POLICY STATUS. If MOTOR POLICY entities can ever be in two states simultaneously or if there can be gaps between leaving one POLICY STATUS and joining another, then two relationships between DATE and POLICY HISTORY would be required, one showing 'start' dates, the other 'end' dates.

So the foregoing deals with the first type of event modelling and we can now move on to the second situation where it is required to record data about the 'event' itself. Consider a retail banking operation. There will be CURRENT ACCOUNTS which over time will have various ACCOUNT MOVEMENTS, for example cheque withdrawals and Bank Giro Credits. Also from time to time it will be necessary to send out a STATEMENT which will cover a number of ACCOUNT MOVEMENTS. The STATEMENT itself is the event. It is necessary to know that one has been despatched and the entries that were on it. This can be modelled as in Figure 3.10. This is obviously highly simplified and there are a number of difficulties with the structure as shown: STATEMENTS are obviously directly related to ACCOUNTS and are themselves DATE dependent. Also there is a clear requirement to be able to duplicate or repeat STATEMENTS which this model does not allow. But the important point here is that the event of producing a STATEMENT becomes an entity type. It represents something of importance, something that needs to be recorded.

It is necessary here to put in a word of warning. Part 3 of this book is concerned with process analysis and 'events' are integral to the techniques expounded there. The events identified at that point do not automatically appear as entity types in the data model. The concern here is with those events that need to be recorded or which fundamentally affect the data that is in the model.

It is important to ensure that these events are adequately modelled. If this is not done, inadequate history of 'what happened when' will result and it will be difficult to place the results of enquiries into the right perspective. For most major entity types some form of event recording will be necessary. It should only be omitted, therefore, once the data analyst is entirely satisfied that it is not needed.

Modelling Hierarchies

A hierarchical structure is often presented to the analyst at a superficial level. This can result in highly restrictive and inflexible models. The underlying structure has to be explored.

Hierarchial structures are fairly common in real-world situations. They typically occur in company organisations or business facility breakdowns, but many other hierarchies are possible. As distinct from categorisations, hierarchies are likely where something of interest can be broken down into a number of lower level items which themselves may be broken down, etc, and where all the lower level items may be regarded as being part of the higher level. Let us take an example of a company structure and demonstrate the hierarchial nature of the structure, as in Figure 3.11. A data analyst could easily be in a situation where this is how the company organisation is explained and a data model such as that shown could result.

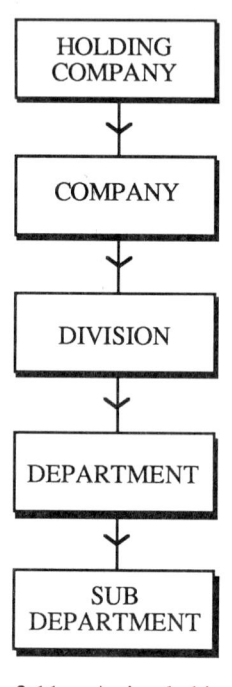

```
┌─────────────┐
│  HOLDING    │
│  COMPANY    │
└──────┬──────┘
       ↓
┌─────────────┐
│  COMPANY    │
└──────┬──────┘
       ↓
┌─────────────┐
│  DIVISION   │
└──────┬──────┘
       ↓
┌─────────────┐
│ DEPARTMENT  │
└──────┬──────┘
       ↓
┌─────────────┐
│    SUB      │
│ DEPARTMENT  │
└─────────────┘
```

Figure 3.11 A simple hierarchy

When examining this structure, however, there is one immediate problem. Where the entity types named in the structure change with the level in the hierarchy, the structure becomes very inflexible. In this example, this is clearly shown. Start by taking entity type HOLDING COMPANY, of which there is, presumably, only one occurrence. There is also an entity type COMPANY of which there will be many occurrences, each of which must report to the HOLDING COMPANY. Each COMPANY may be split into DIVISIONS and each DIVISION must be directly related to one COMPANY. This can be repeated for DEPARTMENT and SUB DEPARTMENT. But what about a small company within the structure? According to the hierarchy, if the COMPANY has DEPARTMENTS then it must also be split into DIVISIONS as a DEPARTMENT does not relate to COMPANY directly.

Other restrictions may also be built into the structure, each causing more and more inflexibility. The problems regarding modelling company structures are quite complex and will be treated in a later section of this chapter, but the problem with hierarchies is well demonstrated by the example. Where a hierarchy of this type seems necessary, to accept the structure as given can cause later difficulties as all actual occurrences of the hierarchy must conform to the constraints of the structure.

In these situations the data analyst must be very aware that the actual hierarchy may be more variable than that which is stated explicitly. There is no shortcut to solving this problem apart from questioning the information given and trying to fit the 'actual' to the 'stated' to detect where there may be a mismatch. It is necessary to search actively for anomalies as it is behind these that the truth often lies.

Above all it is necessary to question whether the entity types in the hierarchy are the same or different. This is the key to the solution of the problem. There are two typical solutions; the choice depends on how homogeneous the hierarchy is. Both solutions will be examined in the next section.

Recursive Structures

Recursive relationships can be resolved by interposing an additional entity type which effectively represents the relationship itself. The entity type can also be used to hold additional data about the relationship occurrence and enables the relationship to be qualified where multiple recursive relationships exist.

We have already met the Recursive Relationship in Chapter 2 and the point was made that this relationship typically occurs in Bill-of-Material operations. This is a particular form of hierarchy. In B-O-M structures this is, typically, a PART entity type. But a particular occurrence of PART may be used to construct one or several other PARTS. Furthermore, our original PART may be constructed from yet more PARTS. Thus, PART has a true Many-to-Many Relationship with itself. A PART can be used to construct many PARTS and a PART can be constructed from many PARTS. The hierarchy, is multi-level and will vary in depth from one occurrence of the hierarchy to another, depending on the complexity of the actual PARTS involved. The varying depth is the major problem and prevents the use of a simple hierarchy to reflect this structure. But there are also two other difficulties. Firstly, there is a Many-to-Many Relationship to resolve. This is not difficult as some other entity type can be interposed in the middle but, even if we had a simple One-to-Many Recursive Relationship, there would still be a problem. It is not normally permissible to have occurrences of an entity type directly related to other occurrences of the same entity type. It is not always supported by an implementation and it is one of the few times when implementation factors can affect the model.

So we need to 'invent' something to put into the model. It is needed to resolve the Recursive Relationship itself and also to breakdown the Many-to-Many element of B-O-M applications.

What should this entity type be? We need to examine the meaning of the relationship. At the start of the section the meaning of the structure was described in terms of usage: *how* a PART is used and *what* a PART uses. These are the two One-to-Many Relationships resulting from the Many-to-Many and the intervening entity type simply qualifies an occurrence of the relationship. Figure 3.12 helps make this clear. The USAGE entity type simply consists of two Part Numbers, one supporting each relationship and both functioning as Identifying Attributes. But how does this work? The USAGE entity type links two occurrences of PART together and states the nature of that linkage. Starting at an occurrence of PART the 'uses' relationship can find a number of USAGE entity occurrences. For each USAGE entity occurrence so found, the 'used by'

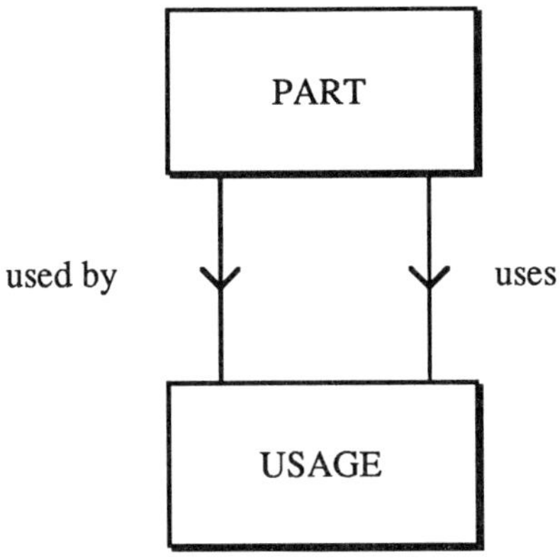
PART
used by
uses
USAGE

PART [Part Number, Part Description, etc]

USAGE [Used by Part Number, Uses Part Number]

Figure 3.12 A Resolved Recursive Relationship

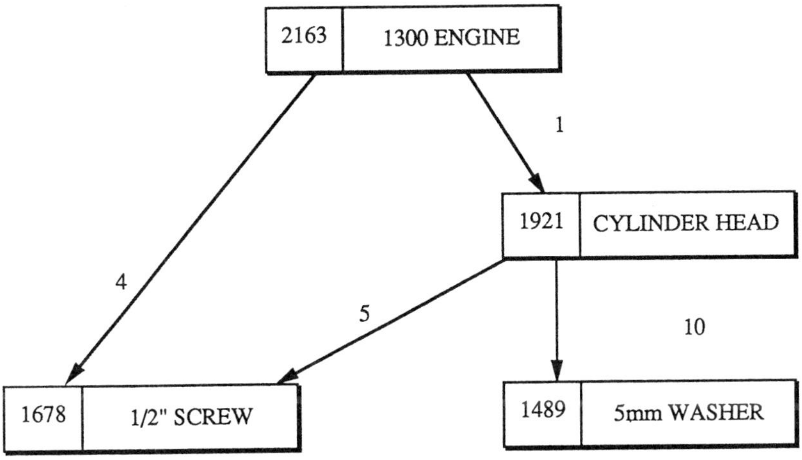
2163
1300 ENGINE
1
1921
CYLINDER HEAD
4
5
10
1678
1/2" SCREW
1489
5mm WASHER

Figure 3.13 Part of a B-O-M structure which can be represented via a structure such as Figure 3.12

relationship will point to one, and only one, other occurrence of PART. Each of these will be the PARTS that are used to construct the PART we started with. Starting at the same point it is possible to discover how a given PART is used by manipulating the relationships the other way round. The 'used by' relationship will give a number of USAGE entities and each of these will be related to one, and only one, PART via the 'uses' relationship.

If you are now feeling confused, don't worry! Take a break and think it through .... it does work!

Examine the example given in Figure 3.13. This shows part of a B-O-M structure which might apply to a Car Manufacturer. Only a few examples of PART are shown, but already an illustration of the type of difficulty that we are attempting to model can be seen. Each PART may be used to make other PARTS and so on.

The depth of the hierarchy will vary. Any questions such as "Which parts use a 1/2 inch screw?" or "What parts go to make up a 1300 engine?" will need to be answered. Figure 3.13 shows not only which PARTS are used, but how many.

This would need to be incorporated into Figure 3.12 as an additional attribute of the USAGE entity type. This then enables us to answer questions such as "How many 1/2 inch screws are needed to make a 1300 engine?". The answer of course is not four, as it is necessary to include those used to construct subsidiary PARTS. The computer application, therefore would have to track down the hierarchy to calculate the correct answer. In Figure 3.13 the answer is nine, but this would only be a small section of the B-O-M structure for any application.

This sort of solution gives a flexible structure. Just about any form of Recursive Relationship can be modelled this way and most situations can be handled. The structure is inherently flexible but, as with all data modelling techniques, the more flexible the solution, the harder it is actually to control the implementation of the structure. Somebody, somewhere is going to have the responsibility to maintain this flexibility and control which occurrences of the entity type relate to which others. The structure, must be clearly understood, not just by the data analyst but by the relevant users as well, otherwise chaos can result. Similarly, consideration must be given to reporting on what structures have been set up in order to monitor and control their usage. It is all too tempting to use such structures as an 'escape route' to avoid doing thorough analysis. Suffice it to say that although this type of structure gives great flexibility, it should not be used without due consideration.

One further aspect of Recursive Relationships occurs where the relationship itself can signify different things. Assume that there is an application that holds details of various CUSTOMERS. It may also be of interest to the organisation to know which CUSTOMERS are related in some way and this is a typical recursive structure, where one occurrence of CUSTOMER may be related to several others. What happens if it is also useful to know what form the relationship takes? For example, some CUSTOMERS may be subsidiaries of others, some may be jointly owned by two or more other CUSTOMERS, or the grouping may be imposed by the organisation simply because it wishes to group those CUSTOMERS together for some reason. This is not really any different from many other data modelling situations. The nature of the relationship, which is now represented by the USAGE entity type, can be classified and therefore the USAGE entity type itself can be categorised or qualified. Figure 3.14 shows this structure. All that has been done is to add this qualification but it does give some advantages. A given CUSTOMER may if necessary, be related to another particular CUSTOMER several times, each CUSTOMER GROUPING entity occurrence being qualified by the CUSTOMER GROUPING TYPE – Subsidiary, Jointly owned or Imposed, to take the above example. Obviously this allows a CUSTOMER to participate in a number of CUSTOMER GROUPINGS which may or may not totally or partially overlap each other. Control over what types of CUSTOMER GROUPING can be set up has also been tightened. Any two CUSTOMERS can only be related via a CUSTOMER GROUPING if that CUSTOMER GROUPING fits one of the previously defined CUSTOMER GROUPING TYPE entity occurrences. If that were not the case, we would have to set up a new occurrence of CUSTOMER GROUPING TYPE *before* the relationship between the CUSTOMERS could be recorded.

It also gives the ability to enquire not only about which CUSTOMERS are related but in what way: "which CUSTOMERS are subsidiaries of CUSTOMER 1894?" or "which CUSTOMERS jointly own CUSTOMER 2796?" Again we have increased the usefulness of the underlying relationship.

This basically concludes the discussion on Recursive Relationships, although there are various further developments that can be applied. The concepts required here become more and more abstract in nature and in order to keep in touch with reality, these will be discussed in terms of their most usual application, company organisations. This follows in the next section.

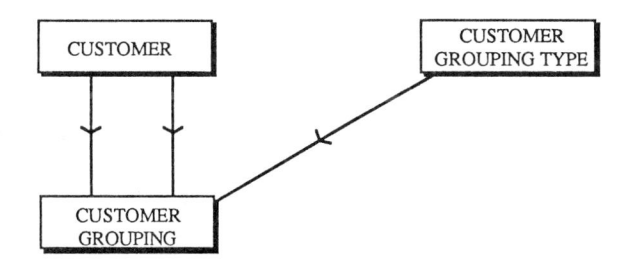

Figure 3.14   A decomposed Recursive Relationship further qualified by type of relationship

Basic Data Analysis Techniques

Company Organisations

Company Organisations are a good example of complex hierarchies. A hierarchical structure can be very restrictive if cast in simple entity types. It can be improved by the following techniques:

- *moving to a more abstract concept for the key entity types such as ORGANISATION UNIT (Figure 3.16);*
- *resolving the resulting recursive relationship in the standard manner (Figure 3.17);*
- *realising that there may be several hierachies embedded within a structure;*
- *distinguishing between the structure of the units, and their functional grouping, thus giving rise to two hierachical structures which are interrelated (Figure 3.19).*

Why pick on Company Organisation as a separate section of its own? There are two reasons: firstly, it is a convenient vehicle to convey the concepts that should be discussed at this point; secondly, this area is not often considered as an area for analysis but there are good reasons for including it.

So why should the company structure be modelled? Part of our underlying philosophy for using data analysis is that it is necessary to structure and control the data that is of importance to the organisation. Companies will always structure their business and the information they use will be related to this structure. If we are trying to control the data we hold, this data must be related to the structure. If Department A can update entity types A,B and C, but not entity type D, then this rule should be modelled in the structure. Furthermore, if we are developing an information processing system that can be used to service enquiries, it is necessary to put the results of those enquiries into context. 'Who can do what' and 'who has done what' are valid questions and, unless there is a clearly defined structure to relate the data to, these questions cannot adequately be answered. This goes right back to the GIGO argument. GIGO occurs because traditional systems have tended *not* to prevent people from inputting data or initiating processes that are not permissible. The details of the way the organisation is structured and the rules that have been imposed (normally for very good business reasons) have normally not been held in computer systems.

A number of major problems will be encountered as soon as this area is tackled. Figure 3.15 shows a typical organisation hierarchy. There will be one or more COMPANIES at the top of the hierarchy and within them there are DIVISIONS, DEPARTMENTS and SUB-DEPARTMENTS. This is the typical inflexible hierarchy. As far as this model goes, it states, for example,

```
┌─────────┐
│ COMPANY │
└────┬────┘
     ▼
┌──────────┐
│ DIVISION │
└────┬─────┘
     ▼
┌────────────┐
│ DEPARTMENT │
└─────┬──────┘
      ▼
┌────────────┐
│    SUB     │
│ DEPARTMENT │
└────────────┘
```

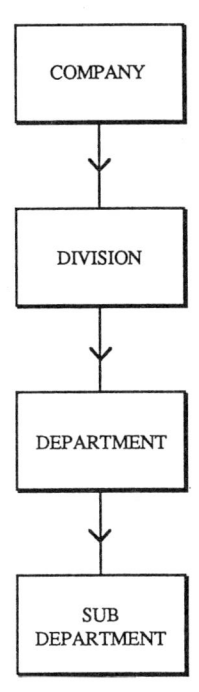

Figure 3.15   A simple organisation hierarchy

```
┌──────────────┐
│ ORGANISATION │
│     UNIT     │↩
└──────────────┘
```

Figure 3.16   An abstracted Company Organisation Structure

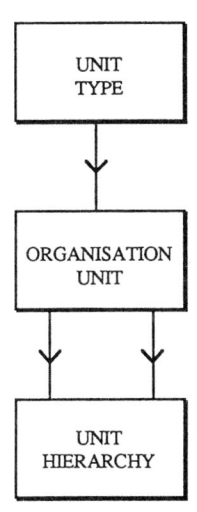

```
┌────────┐
│  UNIT  │
│  TYPE  │
└───┬────┘
    ▼
┌──────────────┐
│ ORGANISATION │
│     UNIT     │
└──┬────────┬──┘
   ▼        ▼
┌────────────┐
│    UNIT    │
│ HIERARCHY  │
└────────────┘
```

Figure 3.17   An organisation structure after being resolved
to remove the Many-to-Many

Basic Data Analysis Techniques

that if any COMPANY has DEPARTMENTS then the COMPANY must itself be split into DIVISIONS. Additionally, the COMPANY must be at the top of the hierarchy. If a DIVISION has responsibility for several subsidiary COMPANIES then this cannot be modelled and, unless we are dealing with a small organisation, it is unlikely that this can be modelled correctly using a simple hierarchy. Another difficulty occurs if there is an occurrence of one entity, perhaps a DEPARTMENT, that has split reporting to two or more DIVISIONS.

It is necessary to ask some particular questions to try and resolve this.Initially, are all these entity types different things or are they the same? If it can be argued that they are much the same – that is, they have common attributes – then we are getting towards the answer. The information needed about a DEPARTMENT is probably much the same as the data needed about a DIVISION, or a COMPANY or a SUB-DEPARTMENT. If this is the case then we have a different type of structure. What we really have is shown in Figure 3.16, which at first sight looks rather simplistic. There is now one entity type, ORGANISATION UNIT, which is recursive in structure. As with all Many-to-Many Recursives this can be resolved as shown in Figure 3.17; this has been done using the techniques previously discussed. An entity type, UNIT TYPE, has been added to help to distinguish which sort of unit we are dealing with, Department, Division or whatever. This is now starting to get somewhere.

It is now possible to have, for example, Companies reporting to Divisions, or Departments reporting directly to Companies. Also any particular ORGANISATION UNIT can report to more than one other ORGANISATION UNIT. Remember as well, that there may be more than one hierarchy in the organisation. If this is true we may need to qualify the entity type UNIT HIERARCHY in order to clarify which hierarchy we are talking about. This could be achieved by introducing an entity type HIERARCHY TYPE or similar.

The foregoing, however, is not really adequate to model the full glory of many organisations. Consider Figure 3.18. We can model the basic reporting structure of the organisation (Figure 3.17), but it is necessary to distinguish between whether the concern is with a part of the structure or with a combination or group of units. If, say the performance of the Sales Division is important, we need to consider not only North and South Regions, Administration, and presumably the salesman reporting here, but also the Sales Department of the Associated Company. Taking this to its logical conclusion it is possible to have a unit with a multiplicity of reporting lines in different hierarchies and additionally being part of a number of groups of ORGANISATION UNITS for different purposes. Obviously, all the ORGANISATION UNITS together are in one big group that represents the whole organisation, so the groups themselves may be hierarchial.

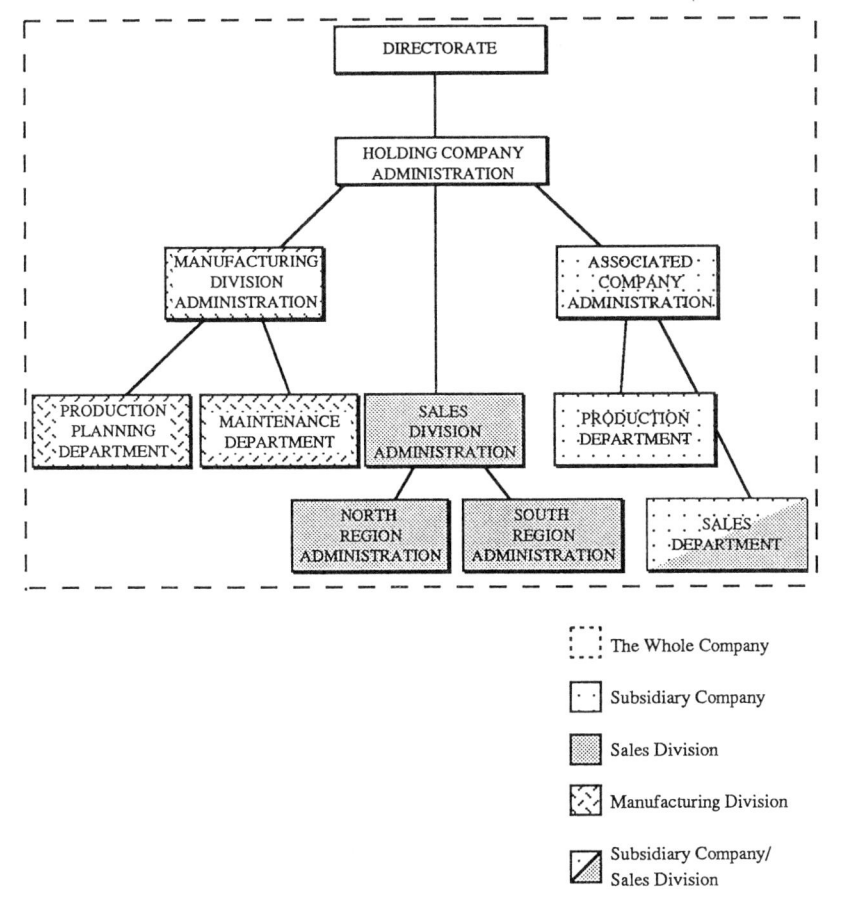

Figure 3.18 Part of an organisation structure showing basic reporting structure plus groups of units with common responsibility

This is now getting rather abstract but it can still be modelled, as shown in Figure 3.19. To clarify the meaning of this it is useful to give some examples of what is meant. ORGANISATION GROUPS would be such things as each Division or Company or large Department. The ORGANISATION UNITS would be the nodes in the structure, for example Divisonal Administration, a Department that is not sub-divided, or the Directorate. This then allows us to relate the things the business does either by the individual ORGANISATION UNIT that caused it to happen or by the group that was responsible for it.

What has been done here is to create an extremely flexible structure that will permit just about any type of organisation structure and allow any type of statistical or management analysis that is required. All we have really done is to create and relate two recursive structures, but it is the creation of the Recursive Relationships that is of interest. We moved from the simple hierarchy and its fairly concrete concepts such as DEPARTMENT to something much more abstract, ORGANISATION UNIT or ORGANISATION GROUP. This is an important point. By looking at the problem from a higher, more abstract level it has been possible to build in much more flexibility and recognise that there is no essential difference between the entity types used in the simple hierarchy.

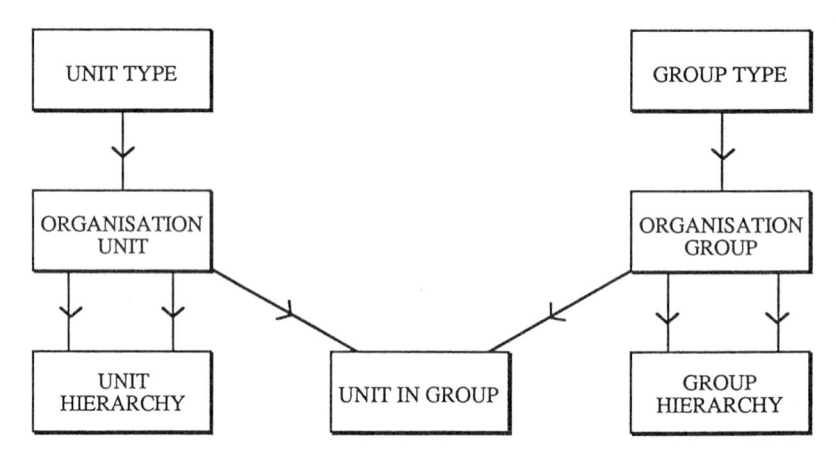

Figure 3.19   A Data Model representing the structure in Figure 3.18

Two warnings should be issued here. Firstly, the data analyst should not rush in and produce such a structure without first determining that it is necessary for correct modelling. As with all modelling, the more flexible the implemented structure, the harder this is to understand and control. We should always aim to produce the simplest structure that will satisfy the requirements. Remember, wherever possible "Keep it Simple, Stupid!"

Secondly, the data analyst should be careful how these concepts are introduced to the relevant user representatives. The concepts are nebulous and should be gradually introduced. It is often necessary to begin with a simple hierarchy and gradually develop the full solution by showing why each degree of abstraction or further sophistication is needed.

This will help to ensure that the data analyst also understands the usefulness of the solution in terms of the particular problem being addressed. A further discussion of the importance of abstract modelling can be found in Chapter 5.

This concludes the full discussions of complex Recursive Relationships and we can now progress to some easier concepts.

Chapter 4

Techniques for Refining the Model

Entity Splitting

By examining the attributes, additional entity types can be created.

- *Repeating groups: new entity type subservient to original;*
- *Variable-length attributes: standardise length and split out;*
- *Code tables: set up an entity type to hold the codes and relate to original;*
- *Indicators: set up a separate entity type to hold values;*
- *Implied events: break out attributes that change over time and create a new entity type;*
- *Optional attributes: break out these attributes but review regularly.*

We can now move on to the second group of techniques which are concerned with improving a model and ensuring that the right structure has been established. An experienced data analyst will apply these techniques automatically during the initial phase of model building but it is useful to review all models against these techniques to check for correctness of design.

One of the basic premises of the approach is that of breaking data into the smallest meaningful chunks that are viable. Obviously this will require that each entity type be examined to determine if it should be broken down still further. Typically, we are looking for the following situations: repeating groups of attributes or variable length attributes embedded in entity types, code tables, indicators, categorisation rules and implied events.

First of all then, let us examine repeating groups of attributes and variable length attributes. In the approach taken here, entity types must be of fixed size and repeating groups or variable-length attributes need to be catered for some other way, normally by creating more entity types. We'll stick with the motor insurance example.

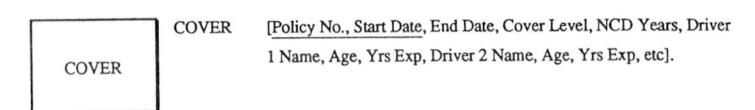

```
┌─────────┐      COVER     [Policy No., Start Date, End Date, Cover Level, NCD Years, Driver
│  COVER  │                 1 Name, Age, Yrs Exp, Driver 2 Name, Age, Yrs Exp, etc].
└─────────┘
```

Figure 4.1   Example of a Repeating Group

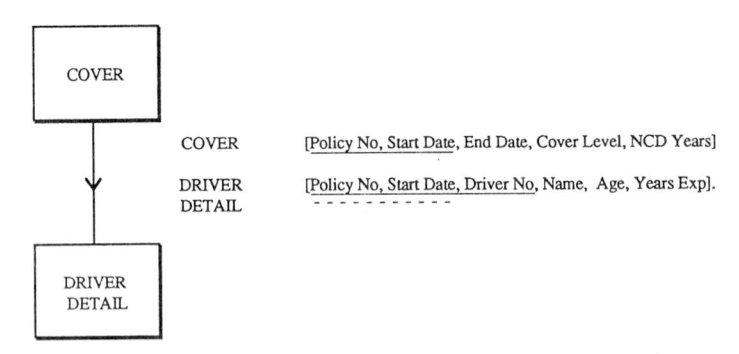

```
┌─────────┐
│  COVER  │
└────┬────┘
     │          COVER         [Policy No, Start Date, End Date, Cover Level, NCD Years]
     ▼
                DRIVER        [Policy No, Start Date, Driver No, Name, Age, Years Exp].
                DETAIL         - - - - - - - - - -
┌─────────┐
│ DRIVER  │
│ DETAIL  │
└─────────┘
```

Figure 4.2   Resolution of the repeating group into two entity types

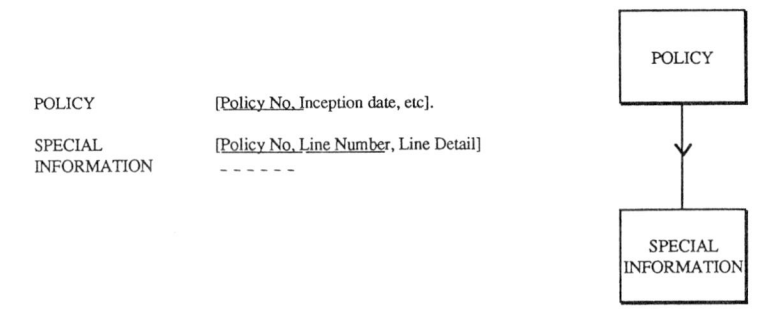

```
                                                          ┌─────────┐
                                                          │ POLICY  │
                                                          └────┬────┘
POLICY              [Policy No, Inception date, etc].          │
                                                               ▼
SPECIAL             [Policy No, Line Number, Line Detail]
INFORMATION          - - - - - -
                                                          ┌──────────┐
                                                          │ SPECIAL  │
                                                          │INFORMATION│
                                                          └──────────┘
```

Figure 4.3   Removing variable length attributes into a separate entity

```
┌─────────┐      VEHICLE   [Reg. No., Make, Model, Engine Size, Type of Body, No. Seats, Right Hand
│ VEHICLE │                 Drive?, Registered Owner, Year Manufactured, Colour, etc].
└─────────┘
```

Figure 4.4   An example of an embedded code table and an indicator

56   Practical Data Analysis

Repeating groups

It is possible to define an entity type called COVER with the attributes shown in Figure 4.1. The problem here is that the number of times the driver group repeats is variable and violates the fixed length rule. This can be resolved as shown in Figure 4.2. Now there may be as many DRIVER DETAILS on COVER as required and each DRIVER DETAIL is tied to the appropriate COVER via the relationship.

Variable-length attributes

It is often the case that a particular entity type will require some descriptive text and the user may not wish for this to be constrained in any way. This variable-length attribute should be removed and placed in a separate entity as shown in Figure 4.3. There can now be as many SPECIAL INFO entities as necessary to hold the free form information about each POLICY. The SPECIAL INFO entity type can be of fixed length.

Code tables and indicators

Referring to Figure 4.4, first of all, the make and model should be immediately split into separate entity types. Passing over that point for the moment, consider the two attributes 'Type of Body', and 'Right Hand Drive?' Taking Type of Body, it is fairly obvious that only certain values would be permissible for this attribute. Such things as Hatchback or Saloon would be OK, Muscular would not (at least not in this model!). Going back to the original principle of GIGO, if we want to limit the range of values in Type of Body, traditionally this would be held as a table somewhere in the program code. However, this is awkward and gives problems when new styles are developed (for example Hatchback is fairly new). Furthermore, if there is a need to analyse by Type of Body for some reason then there is no easy way of doing this. An almost identical argument can be applied to the Right Hand Drive Indicator. The way it is set up here is obviously a Yes/No field or, perhaps more meaningfully, a Right/Left value. Again only two values are allowed and to ensure clean data the permissible values must be made explicit.

The solution to this problem is shown in Figure 4.5. The code table and the indicator have been split into separate entity types. It is now possible to specify in advance what types of body are valid and hold these as occurrences of the entity type. Whenever a

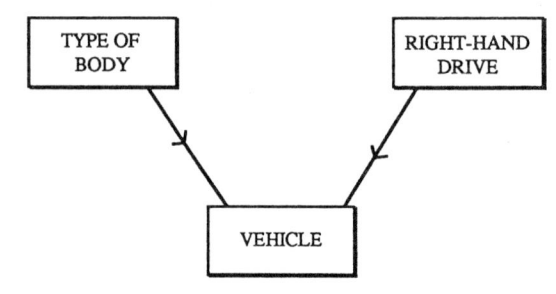

```
        ┌─────────┐                    ┌──────────┐
        │ TYPE OF │                    │RIGHT-HAND│
        │  BODY   │                    │  DRIVE   │
        └────┬────┘                    └────┬─────┘
             │                              │
              \                            /
               \                          /
                \        ┌─────────┐     /
                 _____│ VEHICLE │____/
                         └─────────┘
```

TYPE OF BODY   [Body Type Code, Description]

RIGHT-HAND DRIVE   [RHD Code, Meaning]

VEHICLE   [Reg No., Make, Model, Engine Size, Body Type Code, No. Seats, RHD Code, Registered Owner, Year Manufactured, Colour etc]

Figure 4.5   Splitting out the Code Table and Indicator

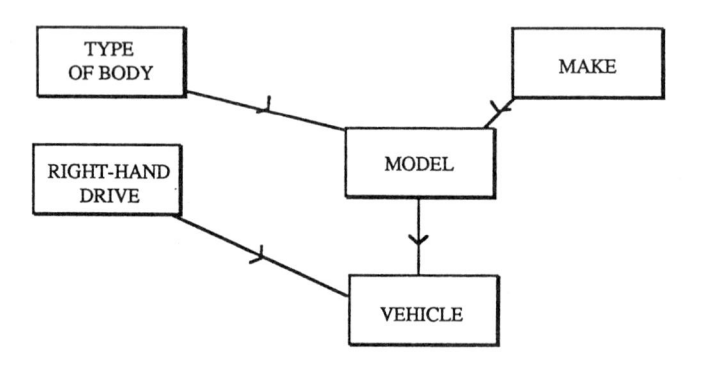

```
        ┌─────────┐                              ┌────────┐
        │  TYPE   │                              │  MAKE  │
        │ OF BODY │                              │        │
        └────┬────┘                              └───┬────┘
             │           ┌─────────┐                 │
             └───────────│  MODEL  │─────────────────┘
        ┌─────────┐      └────┬────┘
        │RIGHT-HAND│          │
        │  DRIVE   │          │
        └────┬─────┘          │
             │     ┌─────────┐│
             └─────│ VEHICLE ├┘
                   └─────────┘
```

TYPE OF BODY   [Body Type Code, Description]

MAKE   [Make Code, Make Name etc]

MODEL   [Make Code, Model Code, Model Name, Engine Size, Body Type Code, No. Seats etc]

RIGHT HAND DRIVE   [RHD Code, Meaning]

VEHICLE   [Reg. No., Make Code, Model Code, RHD Code, Registered Owner, Year Manufactured, Colour etc]

Figure 4.6   Adding Categorisation entity types and the effect on relationships

VEHICLE is stored the system can ensure that it is related to a recognised TYPE OF BODY and a valid value for RIGHT HAND DRIVE. By this measure it is made impossible for a VEHICLE to be stored with a type of body of 'Muscular' and we could also analyse or enquire by type of body should we so wish. Lastly, by giving the user department the ability to add new body types to the list and possibly also delete them as well – although this requires more care – the maintenance problems have been solved. Even though Type of Body may be used by many programs, by making the table explicit it is possible to design the programs in such a way that, even when the values for TYPE OF BODY change, no program changes are necessary and the table will drive the programs correctly.

Categorisation

The next point concerns categorisation. Some of the arguments have already been considered in Chapter 3 and these are not dissimilar to the arguments used for splitting out code tables. Consider Figure 4.6. Make and Model have been taken out as being fairly self-evident but there are two interesting side effects from this enhancement. Firstly, some of the attributes that had previously been held against the VEHICLE can be seen to be attributes of the MODEL, Number of Seats for example. This has tightened still further the constraints applied to the data within the data model. We are now stating that all VEHICLES of a given MODEL will have the same number of seats. This, for example, prevents us from ever having a six-seater Mini.

Secondly, the nature of the relationships has also changed. The TYPE OF BODY entity type now relates to MODEL rather than VEHICLE, again tightening the validity-check and making it more difficult to introduce rubbish data. Not only can there no longer be a Renault 5 with a body type of 'muscular', it *must* be a Renault 5 with a TYPE OF BODY, Hatchback. It is this type of constraint that we should be striving to work into the data model and the data analyst should always consider, when introducing a new entity type, the effects that this may have on existing entity types and relationships.

Implied events

Taking things one stage further, it can be seen that not all the attributes of the entity types we have established will remain unchanged over time. On VEHICLE, two obvious changes may

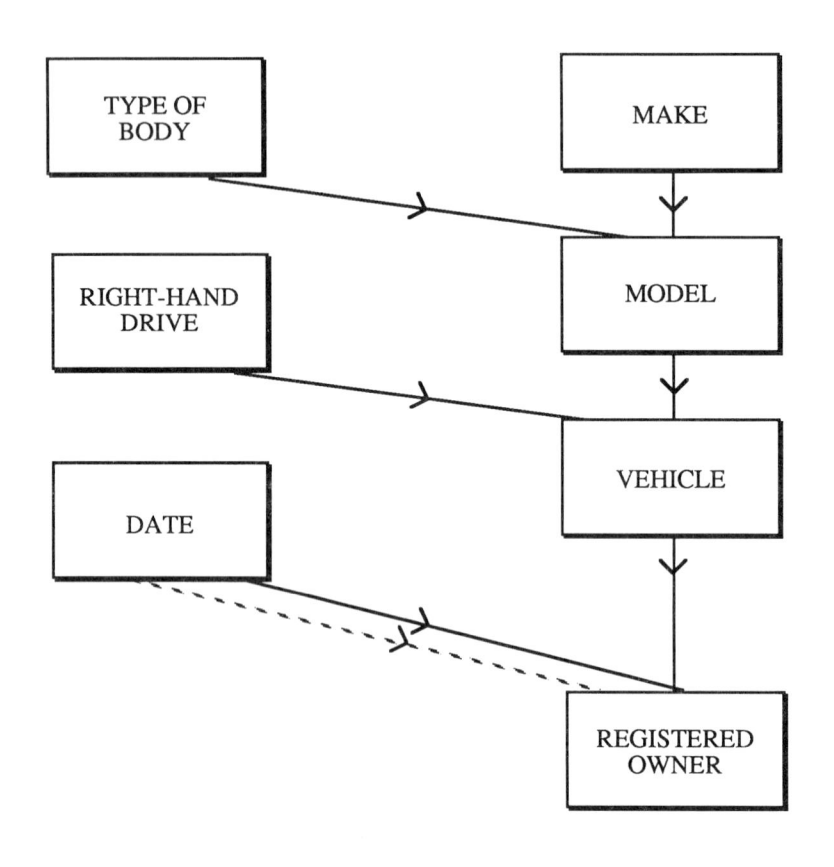

TYPE OF BODY		[Body Type Code, Description].
MAKE		[Make Code, Make Name etc]
MODEL		[Make Code, Model Code, Model Name, Engine Size, Body Type Code, No. Seats, etc]
RIGHTHAND DRIVE		[RHD Code, Meaning]
VEHICLE		[Reg. No., Make Code, Model Code, RHD Code Year Manufactured, Colour etc]
DATE		[Calendar Date]
REGISTERED OWNER		[Reg. No., Owner Name, Start Date, End Date, etc]

Figure 4.7   Adding an implied event - Time dependency

occur. The Registered Owner will change, or should change, each time the vehicle is bought and sold. Also, the colour may change, especially as the vehicle gets older. It is unlikely that historical information is required concerning colour. Once it has changed, it is likely, but not certain, that the old colour will cease to be important. It is much more likely, however, that details of ownership changes will need to be kept. This is one of those situations where, by examining the attributes and asking questions about the possibility of change, hints about events and time-dependency may be forthcoming. In this case the event is change of ownership and the time-dependency is on the Registered Owner. Figure 4.7 makes this clear. There is no restriction on how many Registered Owners there may be and, as it would be necessary to record specific data such as address about each, it is wise to break the Registered Owner out into a separate entity type. Note that two relationships have been shown to DATE, one for Start Date, the One-to-Many; the other for End Date, the Zero-or-One-to-Many. It is probable that the End Date would not always be known, so it is necessary to show this relationship as optional. Lastly, note that Start Date has been included as an Identifying Attribute for REGISTERED OWNER. This is necessary as it is conceivably possible for the VEHICLE to be registered in the same name more than once in its life. The only way these occurrences of REGISTERED OWNER can be differentiated is by utilising Start Date as an identifier for REGISTERED OWNER.

Optional attributes

There are sometimes situations in data analysis where a number of the attributes allocated to an entity type are optional. The question often asked by the data analyst is "Should I make these a separate entity type or not?" The answer is, I'm afraid, "it depends". If there are significant numbers of this entity type and the optional fields are used in less than say 50 per cent of the cases then it may be as well to *implement* them as a separate record, but in data modelling we are not concerned with implementation issues and the decision is less clear cut. A useful rule-of-thumb here is to consider the effect of these attributes. If they were split out would they affect the relationships in which the original entity type participated? If the answer is "Yes" then there are probably two distinct concepts and it would be useful to create an entity type to contain the appropriate attributes. If not, then they might as well be left where they are. Remember, however, that the existence of significant optional attributes may imply that the data analysis is incomplete; the existence of optional attributes should be reviewed periodically to check the correctness of the decision.

A further development of this concerns logically separate groups of attributes within the same entity type. Where these can be seen to occur then they should be split out as it is inherently wrong to combine different things into the same entity type.

Systems analysts have traditionally looked for *similarities* between data when investigating an area. The data analysis approach allows us to reflect the very real variety of data with which the business has to deal. Rather than creating large monolithic records, normally with many embedded codes, optional fields and the like, this approach attempts to split the data down into its constituent parts and make the structure of these entity types explicit. This is the essence of data analysis, but unfortunately, it can be taken too far to be useful and we need to look, as it were, at the opposite side of the coin.

Entity Combination

Clues to entity combination are:

- *Zero-or-One-to-Many Relationships: where an entity type is at the 'many' end of several of these, the entity types at the 'Zero-or-One' end may need to be combined;*
- *Identical Relationships: entity types participating in the same or similar relationships should be considered for combination;*
- *One-to-One Relationships: consider combination;*
- *Common Attributes: if two entity types have the same or similar attributes they may mean the same thing.*

Having established the basic data model and applied the techniques in the preceding section with regard to entity splitting, we need to ensure that we have not needlessly split up entity types that logically belong together. The reason for this is primarily concerned with ensuring that no artificial structures have been introduced into the data model. Conveniently, there exists a fairly well-defined set of pointers to indicate where this may have happened.

First, consider relationships. Figure 4.8 shows part of the Motor Insurance Model. The model has two entity types, CAR and VAN that have identical relationships. Note the use of Zero-or-One-to-Many Relationships between these two entity types and VEHICLE MODIFICATION. Each VEHICLE MODIFICATION can logically only apply to one occurrence of either CAR or VAN and not both. Where this situation of identical relationships exists then almost certainly the two entity types should be combined. Even where there is merely a close correspondence of relationships there may still be value in combining the entity types. It is important, however, to consider the effect on attributes of combining entity types and this will determine the final decision. The effect of combining the two entity types is shown in Figure. 4.9. As can be seen, the model is much simplified as a result and the overall structure is much cleaner and more readily understood. It has been assumed that there is some way of distinguishing between cars and vans; in this example, VEHICLE TYPE would suffice. There is another effect on the relationships, though, which is perhaps more significant: the two Zero-or-One-to-Many Relationships in Figure 4.8 have now been replaced with one One-to-Many Relationship. Figure 4.9 shows that each VEHICLE MODIFICATION must relate to one and only one VEHICLE. Figure 4.8 implied that a VEHICLE MODIFICATION could be related to either a VAN or a CAR, or neither, or both. This is clearly inadequate but is unavoidable with this convention: if it

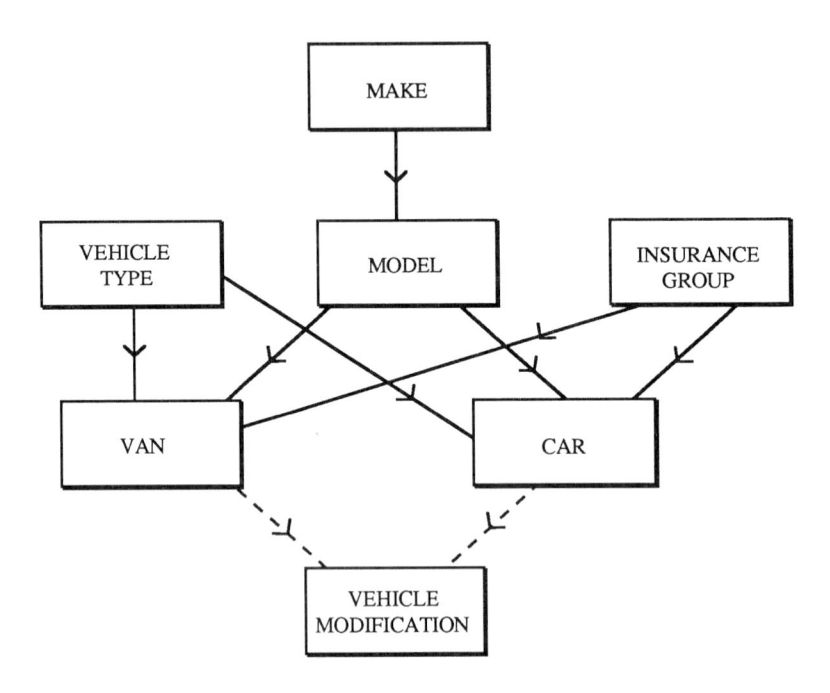

Figure 4.8 A sample Data Model where entity types could be combined

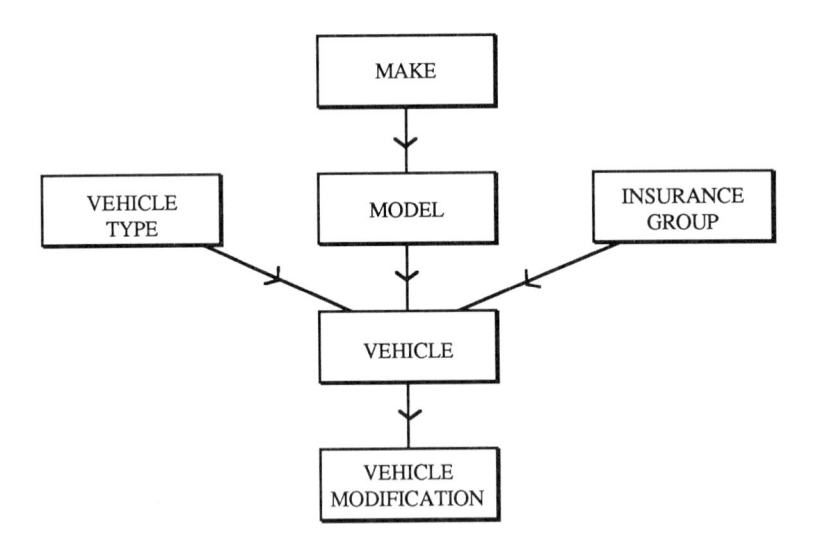

Figure 4.9 The effect of combining two similar entity types

Practical Data Analysis

had been necessary to structure the model in this way, this would need to be included in the back-up documentation. Figure 4.9, however, has resolved this by using the more restrictive One-to-Many and combining the two entity types.

This has highlighted another point concerning relationship-based clues to entity combination. Where an entity type is at the 'many' end of several Zero-or-One-to-Many Relationships it is worth considering if the entity types at the 'Zero-or-One' end of those relationships may be combined. If this is possible it will often, as in Figure 4.9, tighten the constraint on the 'many' entity type and generally yield a cleaner and clearer structure.

The final point on relationships concerns any One-to-One Relationships that may be in the model. Where these occur the analyst should assess the effect on the model of combining the two entity types. If a straight additive combination can be achieved then this is the correct action to take. Sometimes, albeit rarely, such a combination can lead to further complications, especially where either the attributes will not sit easily side by side, or the same attribute occurs in both entity types but would have different values for any two related occurrences or, finally, where the relationships in which the two entity types participate will not easily combine or will be contradictory. Generally speaking, though, the entity types at either end of a One-to-One Relationship should be combined unless there are good reasons for maintaining the separation.

It is not sufficient merely to consider relationships when seeking to combine entity types. It is also necessary to consider the attributes of the entity types, not just in combination with the relationships but also on their own. If two entity types have very similar attributes then these may also be likely candidates for combination – "Have we got the same thing under two different names?" It is normally the case that where there is a close correspondence between the attributes of two entity types they will also share the same or similar relationships, but it is not always the case and close correspondence on attributes or relationships alone may be enough to warrant combination.

Once more, there are dangers in taking this too far. Firstly, there is always the risk of combining two different entity types together even where this is inadvisable thus giving rise to later problems. Secondly, there is the possibility of bending the world to fit the model in the rush to rationalise and simplify the design task. The data analyst must be constantly aware of these dangers and must attempt to balance the techniques outlined in the previous section with the techniques outlined above in an attempt to use each as a check against the other and thus arrive at the correct level of complexity and detail.

Derived Data

Derived data will occur in a number of guises:

- *'fundamental' concepts: these (for example, Account Balance) need to be left in the model to promote acceptance;*
- *existing system data: computer systems often contain significant derived data which should be broken down;*
- *efficiency: data should not be modelled at this stage to assist efficient implementation;*
- *audit and control: unless these requirements give rise to data of their own, they should not be modelled.*

It is central to this philosophy of data analysis that *when a certain piece of data is modelled we should include that data once, and once only. Not just this, but also, if the value of an attribute in an entity type can be calculated or otherwise determined by examining other attributes in other entity types, then this data can be viewed as 'derived data' and should not be modelled.* In other words, "if you've got the bits you can sum them". Remember, in data modelling no account is taken of eventual operational efficiency nor with how long data will remain available to the user. As far as data modelling is concerned, all data, once captured, will be available forever.

This is the 'pure' data analysis viewpoint on derived data. However, in practical terms it is often less easy to ignore this type of data than it may at first appear. In practice it is sometimes necessary to move away, albeit cautiously, from the theoretical standpoint. Some guidance follows on the sorts of problems that can arise.

Firstly, it is necessary to return once more to traditional computer systems and their drawbacks. It is quite likely that the data analyst will often be attempting to develop a data model of an area that is already the subject of a computerised system. Commonly, derived data forms a considerable part of the content of these systems and all attempts should be made to break down this derived data and to model only the 'raw' data from which the field is determined.

This is not always easy to do. Once data has been included in a system, it seems to acquire a reality of its own and it can be difficult to get other interested parties to accept its redundancy.

Secondly, it is necessary to ask if the particular piece of data under scrutiny represents something real in the minds of the users. For example, if one refers to the 'Balance' on an account, most people realise what this represents and would be quite happy, in an everyday sense, with the 'reality' of the concept. But in strict data modelling terms, the balance on an account is derived data – it can always be calculated by summation of all the movements that have occurred across that account. This may be strictly true, but it would be impractical in any implemented system not to include a 'Balance' as all historical data would not necessarily be available. Furthermore, removing the concept of balance from the data model could attract a considerable amount of user-resistance against not including something so 'real'. The data analyst must tread carefully, therefore, and may need to include some items of derived data to circumvent this problem and prevent discussions getting bogged down in esoteric argument about whether or not the concept is real. Include it; but remember that it has been done and, very importantly, why it has been done.

Other arguments in favour of derived data will range round, for example, the convenience of holding some derived data for which frequent and speedy retrieval may be required. Again, although this is very important as a factor in constructing a good implemented system, it is, once more, irrelevant from the viewpoint of the data analyst.

In a similar vein, the data analyst will encounter various Audit and Control requirements and will often be asked to ensure that these requirements can be met. The stated requirements will certainly be of interest and must be satisfied if the end result is to be trusted and relied upon by the users. But the data analyst must ask, once more, if these requirements are relevant to the data model or to the implementation – the test of 'reality' again. In many instances, but not all, this type of requirement will be irrelevant to the data model. These requirements should be noted for later use and not included in the data model.

Coding Structures

> *Code tables present their own set of special problems. These are:*
>
> - *changes over time: code values often change over time and this situation should be resolved with a cross-reference entity type to record the changes;*
> - *combined codes: often several coding structures may be embedded within an existing code table. These variables need to be carefully separated;*
> - *code extensions: in some cases, a cleaner structure can be obtained by extending the list of code values. This should be done if it adds clarity, but the extension must be fully debated before it is accepted.*

Some of the important factors relating to coding structures have already been discussed. However, this aspect of data modelling is very important and even at the risk of being repetitive these factors must be reviewed. Code tables occur, either explicitly or implicitly, whenever some form of shorthand is used to represent a number of related conditions. Common examples would be status settings or transaction type codes.

Why consider coding structures as so important? Codes carry a lot of meaning and are a very useful shorthand for classifying and analysing information. In order to use codes to their full advantage (and there are very few systems that are devoid of codes) then several things must be done. The coding structures used must be controlled and similarly the base data that relates to these codes must also be controlled. With most codes there is a finite number of values that are acceptable as the code. Data analysis provides the opportunity to make these codes explicit and to ensure that only valid values for the code may be input to the system. This is the GIGO reasoning again. If the code can be controlled then the data entering the system can be controlled and it is possible to ensure that any analysis required on particular code values will be correct and more efficient. This also simplifies the programming task. There will be no need to worry, at the program level, about what values are valid or to maintain this list often in several programs or routines. All that needs to be done in the program is to test if the value for the code exists. If it does, it is valid, If it does not, it is invalid. This is obviously much more flexible.

Making code tables visible in the data model, and probably also in the database, draws attention to the code and promotes discussion about its use, the range of values and its underlying meaning. It forces people to consider what the code means and ensures that any grey areas are exposed. This

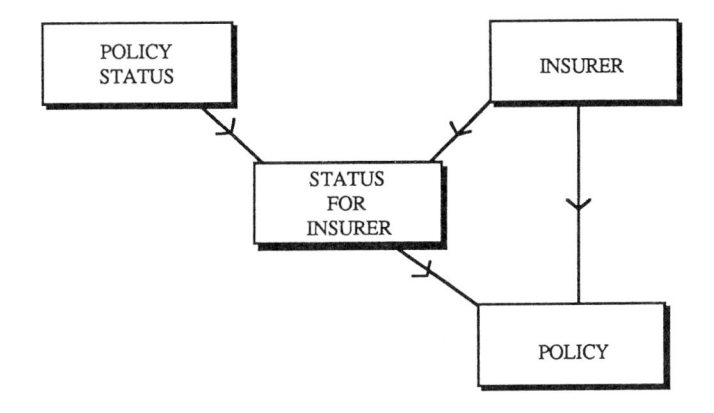

Figure 4.10   No history can be recorded with this structure

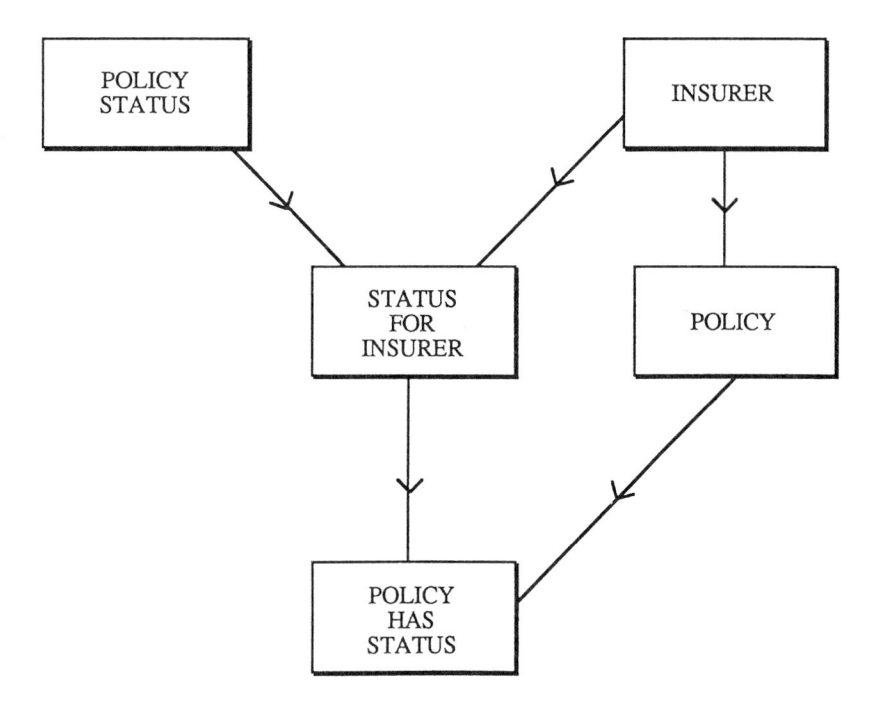

Figure 4.11   An enhanced structure for recording history

Techniques for Refining the Model

visibility of the code table also enables the code to be maintained by the users thus freeing data processing resources. If a new code needs to be added to an existing table because of a change in the business then this, of itself, should be a straightforward operation. If control of the table is given back to the user department then this change can literally take a few seconds to achieve and can be transparent to the data processing department.

One very important use of codes is to reflect the progress of something, or other changes that will occur with time. Typically, codes such as 'status' or 'stage' or 'mode' indicate such usage. It is very useful to be able to control and monitor these changes and the code table itself is a convenient way of doing this. Remember, though, that if the value of a particular code in relation to an occurrence of a different entity type can change over time, then this should be brought out on the model with a suitable entity type interposed to resolve the resultant Many-to-Many Relationship. For example, the structure in Figure 4.10 shows only one POLICY STATUS applicable to a POLICY. Assuming that the POLICY STATUS will change from time to time (Adjustment Outstanding, Renewal Invited, etc) then this structure is inadequate. The structure in Figure 4.11 allows time-stamping of each status change to be recorded. This allows not only accurate reports of which code values apply to which occurrences of the other entity type, but also when it happened and, if necessary, the history can be held to give a complete picture of events.

There are a number of factors which one should bear in mind, especially when extracting codes from existing computerised systems. The most common of these is that several codes are combined together in some way to make use of spare values in the original code table or simply because it "seemed a good idea at the time".

The data analyst needs to be very aware of this problem. Often the combined codes can be identified as being totally independent and they will have been combined purely for expediency. In this case, the codes can be split into separate entity types and 'set free' from the artificial restriction. Equally often, however, the codes will have been combined because there is some relationship between the values of one code and values of the other. If this is the case, careful examination is required in order to extract the full glory of the relationship. The use of 'cross-reference' entity types is discussed in the next section, but this may not be the requirement here. All that may be needed is one code classifying occurrences of the other in a straightforward manner. Each of these situations must be resolved on its own merits and one should remember to tread carefully! Lastly, on this question of separation of codes, possibly the worst situation, and one to send many a stable analyst running for the door, is the bit-coded field where each bit represents a different Yes/No switch and yet all are related in an intricate way. It does happen and the underlying rules can be difficult to extract.

When a system is first implemented, be it a computer system or manual, the code tables are often carefully designed and the values set to reflect the needs of the business.

As the system matures, the code values will often be manipulated and extended in ways which are foreign to the original intentions of the designer. New codes may be added in an uncontrolled manner or existing codes may change their meaning. These can be controlled more easily given good data analysis and clear and easy user-driven maintenance procedures; sometimes an existing code is used in a non-standard way because it is a reasonable fit, but will still need exception processing where the fit is less good. The data analyst must be aware of these exceptions; they suggest that the code has been misused and that there is a need to separate out these exceptions or to extend the range of values that currently exist. Just a drop of psychology here: research in logical thinking shows that negative information is much more meaningful than positive information. It is, though, much harder to get to grips with. Exceptions fit into this theory very well. It can take considerable effort to get to the bottom of exceptions in codes and to uncover the deeper meaning implied by them, but the exceptions will normally say a lot about the limits on the existing code table and shed much light on the main factors originally underlying the code. Don't be put off by exceptions. Remember the 80:20 rule. Eighty per cent of the effort will be expended on 20 per cent of the work and vice versa.

This brings us to another point related to the above. Don't be afraid of extending the code if you can see powerful arguments for so doing, but don't sneak in the change either! Justify your reasons and make sure that your arguments are well understood and get agreement on the change. There can be a lot of resistance to changes to code structures that have existed since the year dot and with which everyone but yourself feels at home with and accepts, warts and all!

Why would one wish to change a code? Consider the real-life example in Figure. 4.12. In this situation, ACCOUNTS could be either External or Internal and would belong to one ACCOUNT TYPE. In the model as shown, any ACCOUNT could be External or Internal regardless of ACCOUNT TYPE but this is not the case. On examination, only one ACCOUNT TYPE, Current, could have both External and Internal ACCOUNTS. For all other ACCOUNT TYPES, all the ACCOUNTS would be either Internal or External but not a mixture. The solution for this is shown in Figure. 4.13. By extending the ACCOUNT TYPE code table to distinguish between Internal Current Accounts and External Current Accounts then a cleaner solution with more meaning results. Now all Accounts of any particular ACCOUNT TYPE would be either External or Internal, more control can be exercised over data input and the chance of making an exception into a general rule has been avoided. This can be a useful exercise and give tangible benefits but you *must* make sure that it is acceptable to your users.

Finally, there is yet another thing to be aware of on coding structures. When the area being modelled is relatively large and many parts of the business are under scrutiny, the data analyst has to look for the same code being used in different ways in different parts of the organisation. This can take several forms. Firstly, the same coding structure can be encountered under two entirely different names, and more so, the values being used may be different but have the same meaning. Two solutions present themselves. Preferably the aim should be to standardise on one set of values but this, again, must be done with the full cooperation of all involved. It is not the role of the data analyst to dictate to an organisation the rules it should have. If this can not be achieved, then the second option is to carry both code values in the same table and allow access on either code when the system is designed. Which code is used as a unique identifier is irrelevant to the users and a fairly academic choice must be made. It is the ability to gain access via either code which is important.

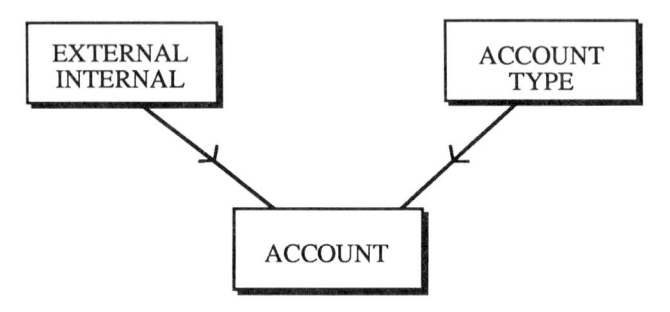

Figure 4.12   Two codes classifying ACCOUNT

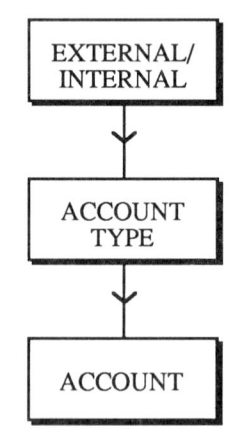

Figure 4.13   A cleaner structure obtained by extending one code

Secondly, different codes may occur using the same name. This is more problematic. One must change and it is up to the Data Analyst to obtain agreement on this issue. The main danger here is the fact that the codes *are* different will not be noticed, with consequent problems later in the development cycle. Unfortunately, we live in an imperfect world and the more common finding is that two different but ovelapping codes are in use where some values agree and others differ. There is no easy way out of this apart from careful analysis and discussion with all interested parties, in order to establish the underlying meaning and use of the codes, gradually working towards an agreed solution.

Cross-Reference Entities

Where two coding structures interrelate to a significant degree, a cross-reference entity type can be used to model the overlap. There are some special points to note:

* *it can be used to constrain the data at a lower level;*
* *it is a powerful way of representing 'valid combinations' of the code;*
* *beware time-dependency.*

The previous section discussed at some length the advantages of breaking out code tables into separate entity types so that they became explicit and visible within the data structure. What this all boils down to is an attempt to specify the rules which raw data must satisfy before it can be entered. There is one very important aspect of rules, however, which has hardly been mentioned so far yet which is very prevalent in the real world.

This is concerned with the interrelationship between different code tables or categorising entity types within the structure. Go back to Figure 4.12 discussed in the last section: this described a situation where there was a little overlap between one rule (EXTERNAL/INTERNAL) and another (ACCOUNT TYPE). Assume, for the sake of argument, that the overlap had been much greater but not complete, that is, a good number of the ACCOUNT TYPES could have both External and Internal ACCOUNTS within them. How can this be made clear in the data model? By taking the solution in Figure 4.13, many extra ACCOUNT TYPES would have to be created in order to achieve our purpose and this could be totally unacceptable. The ACCOUNT TYPE code could carry much meaning for the user and significant changes might prove counterproductive. This type of manipulation could change the very meaning of ACCOUNT TYPE, so that solution is out. It is possible of course, to leave well alone and stick with the model in Figure 4.12, but this just ignores the problem and allows any ACCOUNT TYPE to have a mixture of Internal and External ACCOUNTS.

So a new solution is needed. This is achieved by interposing a 'cross-reference' entity type as shown in Figure 4.14. What are we stating with such a structure? Firstly, this shows that there is some form of interrelationship between ACCOUNT TYPE and EXTERNAL/INTERNAL. Those ACCOUNT TYPES that may have a mixture of ACCOUNTS which are either Internal or External would be related to two occurrences of TYPE MAY BE EXT/INT, one relating to the External occurrence of EXTERNAL/INTERNAL the other relating to the Internal occurrence. Where only one is allowed for an ACCOUNT TYPE then there would only be one occurrence of the cross-reference entity type related to the appropriate EXTERNAL/INTERNAL setting. This is a much more powerful and flexible structure than either Figure 4.12 or Figure 4.13 and is generally of more use. Furthermore, once the appropriate settings of the cross-reference entity type have been established for an ACCOUNT TYPE the appropriate ACCOUNTS can be hung from it. If we need, for example, to extract all External ACCOUNTS of ACCOUNT TYPE 'ABCD' then this is easily and quickly achieved without having to search through all occurrences of 'ABCD' rejecting, on average, 50 per cent of them.

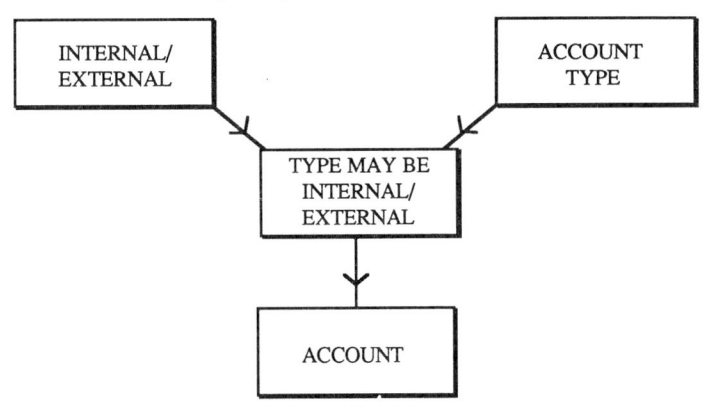

Figure 4.14   A cross-reference entity type between two code tables

Why was time spent discussing altering code structures when this can provide a straightforward solution with no changes to existing structures? The answer comes down to practicalities and benefits. What is important here is the *degree* of overlap between the two code tables. If the two are independent (that is, all ACCOUNT TYPES may have both External and Internal Accounts) then the Model in Figure 4.12 is quite adequate. There is no relationship to bring out and the codes are related only at the actual ACCOUNT level.

This type of solution may also be quite acceptable where the degree of overlap is very small, say less than five per cent as a rule of thumb. Here the additional cost in maintaining and controlling the cross-reference is simply not justifiable in terms of the benefit. This, unfortunately, has to be subjective. If the data is highly sensitive then we may want to include the cross-reference even where the overlap is very slight. The question must be asked, "what are the consequences of getting it wrong?" Again where the overlap is almost total, there may be little value in supporting the cross-reference and a solution as per Figure 4.13 is satisfactory after the appropriate adjustment of the code tables. If the overlap is as much as 95 per cent there is probably no need for the cross-reference but, once more, it is a subjective decision and must be made in the light of individual circumstances.

Techniques for Refining the Model

However, the most likely situation is where the number of 'valid combinations' lies between 10 and 90 per cent of the possible number and in these situations the cross-reference entity type becomes an invaluable mechanism for representing the interrelationship of the codes.

There is one further point concerning situations where two entity types require a cross-reference and where the number of combinations is large and most, but not all, are valid. In the data model it is always advisable to express the positive side of this cross-reference, that is, the *valid* combinations. However, in implementation terms, it may be far more efficient to hold only the exception conditions, that is, *invalid* combinations.

If the cross-reference table is expressed negatively in the data model this can be restrictive as the cross-reference entity type itself cannot be used to constrain occurrences of another entity type. 'What may not be' cannot logically own a relationship to other entity types. So always model the positive side. You may decide to *implement* the negative side, but only if the combination has no other uses.

Finally, a few words of warning and some of encouragement. First, the good news. If, when actively modelling, you suspect that there is a relationship between two codes, put in a cross-reference entity type: it will provoke discussions and highlight the problem. It is quite easy to remove it later if it is felt unnecessary, but if it is not discussed the issue may be ignored completely and a valuable rule of the business, however implicit at the moment, will have been lost. On the warnings side, don't forget the effects of time on the cross-reference or the entity types below it. One or both may be changeable, in which case the cross-reference will have to have this time-dependency brought out in the usual way. Lastly, the cross-reference may be more than just a two-way combination. It may be necessary to feed in three or more tables. This can be handled in exactly the same way. On other occasions, it is possible to get a third code table which has to be cross-referenced with the first cross-reference itself. This is quite permissible if this structure best answers the data requirements.

Role Entities

The use of a 'role' entity type is a very powerful technique for resolving multiple relationships. The advantages are:
- *a cleaner, simpler structure;*
- *avoids a proliferation of Supportive Attributes;*
- *it is easy to add new roles without re-designing.*

This section concerns itself with a very useful technique that can be of great benefit when trying to come to terms with a complex model. Consider the situation as in Figure 4.15. Here there are two fairly obvious entity types, MAN and CAR, and between them there are four Many-to-Many Relationships which have been named here to avoid confusion. The interesting thing about this structure is that each CAR and each MAN occurrence must be able to support all four relationships but it is unlikely that for any given occurrence of either entity type all four relationships would apply. Additionally, the model looks complex because of the many lines that need to be shown. The diagram shows these as Many-to-Many Relationships although One-to-Many and Zero-or-One-to-Many Relationships could equally well apply in other situations.

The use of Many-to-Many Relationships, however, yields a clue as to how to resolve this somewhat confusing structure. A possible solution is shown in Figure 4.16. Here the Many-to-Many problem has been resolved, but things have been taken a stage further. By creating a 'general-purpose' entity type – MAN HAS INTEREST IN CAR – a structure has been produced which can detail the type of interest, drives, repairs, etc, in the interposed entity type itself. This means that it is possible to use the same structure whatever the nature of the interest without creating any additional relationships. Also both MAN and CAR now participate in only one relationship each thus simplifying the structure and lending extra clarity to the model. The new entity type, MAN HAS INTEREST IN CAR, acts in the role required by the particular occurrences of MAN and CAR that it links.

This can be taken a little further. In Figure 4.15 there was a set number of relationships between MAN and CAR that controlled the ways in which the two entity types could be related. In Figure 4.16 this constraint has been lost as the nature of the relationship is presumably held as some form of narrative in MAN HAS INTEREST IN CAR. This is unsatisfactory. As with other forms of categorisation, where there is a finite number of settings for an attribute this should be made explicit. Figure 4.17 details the full structure for this situation. This states that the role must conform to the specific values that correspond to occurrences of TYPE OF INTEREST. The constraint that was lost in the transition from Figure 4.15 to Figure 4.16 has been re-imposed.

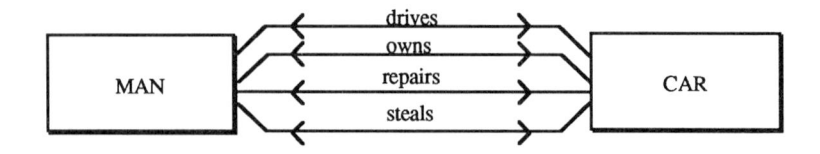

Figure 4.15 A multiplicity of relationships between two entity types

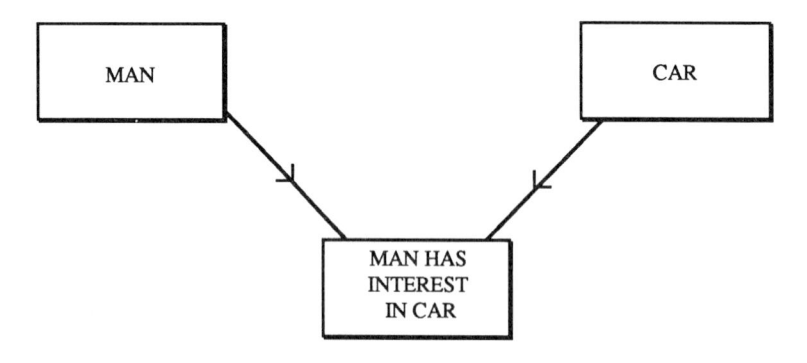

Figure 4.16 A first resolution and creation of an 'in-role' entity type

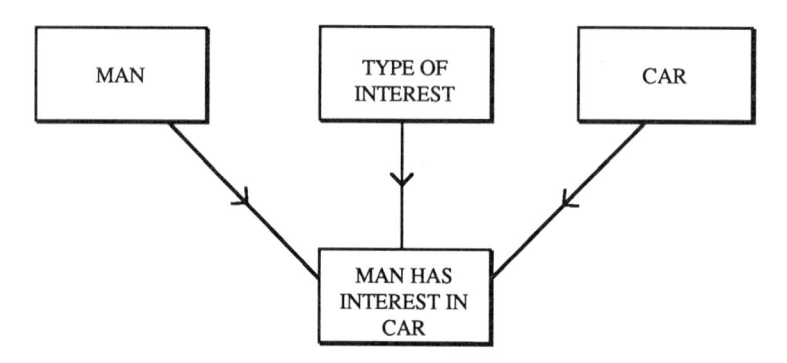

Figure 4.17 Making the role explicit

78 Practical Data Analysis

This technique is of use wherever there are a number of relationships between two entity types, regardless of the type of relationship concerned. The major advantages are twofold. Firstly, it has simplified and clarified the data model and it is no longer necessary to support the capability of each entity type participating in a number of relationships when not all will be required. The simplification of the relationship ensures a more efficient and workable implementation. This also means that there is no need to have a proliferation of Supportive Attributes in the entity types, each attribute referring to occurrences of the same entity type, thus causing confusion and redundancy when storing new data. Secondly, it has been made extremely easy to add new types of role between the entity types without having to include additional relationships and avoiding any change to the structure. If we need to add a new role, let us say 'sells', then this is simply a matter of adding a new occurrence of the entity type TYPE OF INTEREST.

Tightening Constraints

By trying to 'raise' the level at which relationships are expressed, a number of benefits may be obtained:

- *basic data validation is improved;*
- *further rules and entity types can be uncovered;*
- *redundancy can be reduced.*

This section turns attention to ensuring that relationships have been shown at the correct level within the data model. In particular, this concentrates on examining the possibility of raising the level at which a given relationship is shown. The question to be asked is, "can we place this relationship a level higher without losing control over the data?" The goal is to express the rules governing the data at the highest possible level. The higher the level of expression, the greater the level of control exerted over the data and this aids the clarity of the model and avoids redundancy in the implementation. Consider Figure. 4.18. Here there are MAKES, each of which produces a number of MODELS of which there will be a number of VEHICLE occurrences. Furthermore, this states that each VEHICLE must be of a particular VEHICLE TYPE (Private, Heavy Goods, etc). Now there is nothing 'wrong' with this data model and the rules as stated are perfectly correct. But this structure should be questionned to see if it can be improved. Assume that this leads to the discovery that as a MODEL is a fairly specific thing, then each VEHICLE of a MODEL will be of the same VEHICLE TYPE. If this is the case, then the structure shown in Figure 4.19 is more appropriate. We have raised the level of the relationship from VEHICLE TYPE and exerted extra control over the occurrences of VEHICLE. In Figure 4.18 a VEHICLE could be of any VEHICLE TYPE, Figure 4.19 states that a VEHICLE must be of the one VEHICLE TYPE appropriate for the MODEL.

This is an improvement, but we can do better, as shown in Figure 4.20.

We should have tried to place VEHICLE TYPE above MAKE and tested the rule "MAKES can only produce VEHICLES of one VEHICLE TYPE". This is not true, but it has uncovered a further 'rule' about the data. In this case, it shows that a MAKE only produces VEHICLES of certain VEHICLE TYPES.

This has exposed the need for a cross-reference entity type and this is included. This has tightened the rules once more. Not only must a VEHICLE be of the VEHICLE TYPE appropriate to the MODEL, but the VEHICLE TYPES available are constrained to those appropriate to the MAKE.

This is a nice example of a further aspect of data modelling, which is the need to express the general rule as opposed to the particular occurrence. The general rule is that a MAKE may produce VEHICLES of a number of VEHICLE TYPES. But, in particular, a given MODEL will only be of one VEHICLE TYPE. This type of structure is very common. It expresses options at the higher level but precision at the lower level. The difference is between 'what may be' and 'what is'. It should be one of the aims of data analysis to separate the 'rules' from the 'data' and this technique is of great benefit in achieving this.

But what about the other aim of reducing redundancy? This comes to the fore when attempting to relate two separate structures. This is shown in Figure 4.21. The attempt here is to show the correct INTEREST RATE for a given MORTGAGE. The INTEREST RATE will change with time and so a suitable entity type – RATE FOR MORTGAGE – has been inserted to trace the history and record the current situation. However, every time there is an interest rate change for a given INTEREST BASIS, it will be necessary to create a new occurrence of RATE FOR MORTGAGE for each affected MORTGAGE.

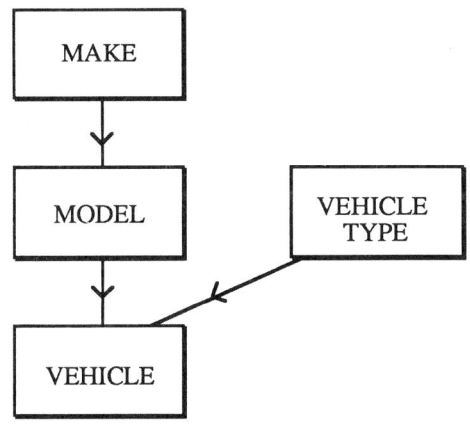

Figure 4.18   A sample structure with 'low level' constraints

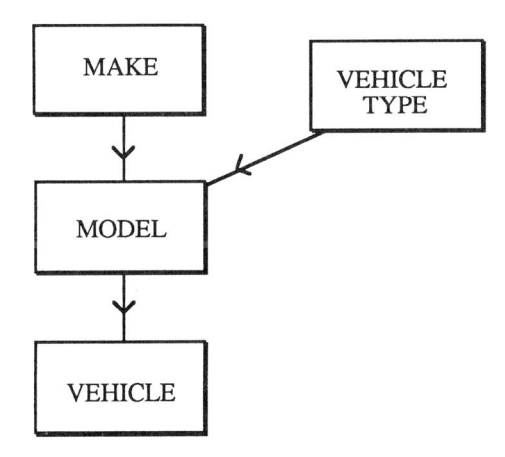

Figure 4.19   The constraint tightened

Techniques for Refining the Model

This is clearly extremely inefficient and also allows the INTEREST BASIS to change during the life of the MORTGAGE, a situation which is almost certainly impossible. The level of this relationship can be altered as shown in Figure 4.22.

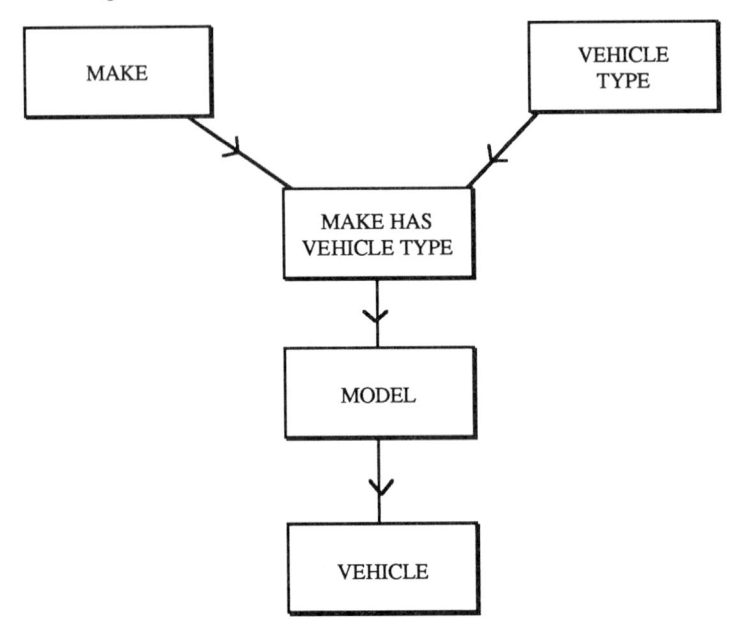

Figure 4.20   A further sophistication illustrating 'general rules' and 'particular occurrences'.

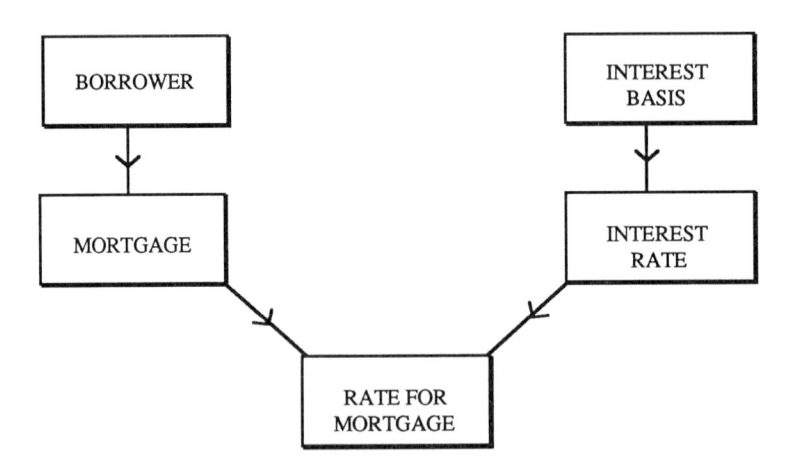

Figure 4.21   An inefficient structure

This shows that the INTEREST BASIS is fixed for the MORTGAGE and that there is no *direct* relationship between MORTGAGE and INTEREST RATE. Care is needed here. If we try to discover which INTEREST RATE applies to a particular MORTGAGE we can get from MORTGAGE to the appropriate INTEREST BASIS easily enough, from the 'many' to the 'one', but we need to ensure that we can select the appropriate INTEREST RATE from the 'many' applicable to the INTEREST BASIS. This can be done *only* if the selection criteria needed are explicit in the enquiry. In this case they are. The enquiry only makes sense in terms of a given point in time and this will be the parameter for selecting the INTEREST RATE. But be warned! If the INTEREST RATE charged can vary on the same date for different MORTGAGES of the same INTEREST BASIS, then this model does not work and would need revision. In this case, we are fairly safe and when the INTEREST RATE does change, all that is needed is to add a new occurrence for the affected INTEREST BASIS and all MORTGAGES will fall into line. Less updating, less duplication, less redundancy.

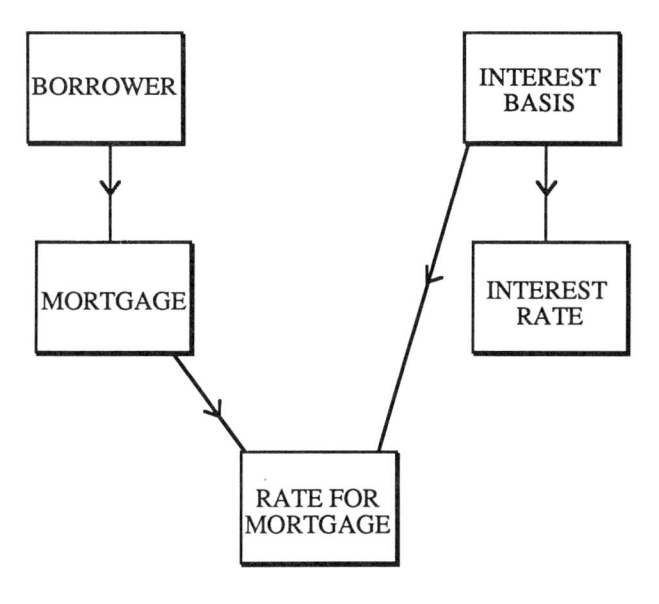

Figure 4.22   Redundancy removed

Techniques for Refining the Model   83

Problems with Updates

Data updates should be examined and two questions asked:

* *can the data change?*
* *is history important?*

This is normally resolved by splitting 'static' and 'volatile' attributes into separate entity types. If this is done then the relationships surrounding the original entity type must be carefully re-examined.

There are very few systems in operation that are 'enquiry' only. The data has to come from somewhere and needs to be kept up-to-date, but there are considerable dangers in the updating process and it is an area where the unsuspecting data analyst can come unstuck. This returns to the necessity to consider time as a factor influencing the structure of the data model and concentrates on answering the question, "What happens when the data in the model changes?".

It is often necessary to preserve history and to reflect the changes that have occurred on the data. Indeed, it can be argued that it is unwise to ever allow data itself to change. All that should happen is that new data, reflecting the new situation, is added to, rather than replacing, the data that is no longer applicable. This is a theoretical argument though, and it is more practical to ask two questions and then act accordingly. Firstly, "Can the data change?" It almost invariably can; there is very little information that, once created, remains forever in a system without some process manipulating it. If perchance, it can be said with confidence that it will not change, then the situation need be considered no longer. However, where change is possible a second question should be asked, "When it changes, do we need to know that it has changed?" The answer again is usually positive. Occasionally there will be no need to refer to the old data and in these circumstances it can simply be replaced. But beware, this is potentially very dangerous as data is being lost without trace and should only be done where the data analyst is certain of the position. Normally the change will need to be recorded and this is usually best done by separating the 'static' data from the 'volatile'.

An example may help to clarify this. Figure 4.23 shows a straightforward extract from the Motor Policy System. We have the POLICY, and for each there will be a number of PAYMENT DUE entity occurrences set up as various sums become due on the POLICY. The POLICY has to reflect all the details relating to the terms of issue, the drivers, level of protection, etc, but many of these details can and will change. The static attributes (Policy Number, Date of Inception, etc) should be separated from the variable

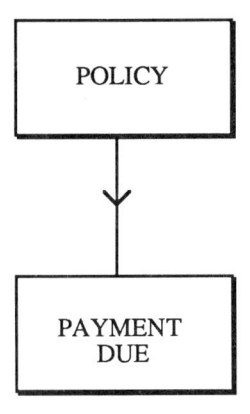

Figure 4.23   The model prior to consideration of updating

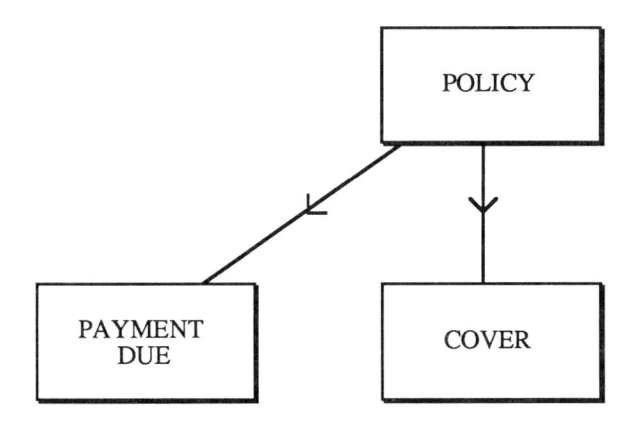

Figure 4.24   The introduction of an 'historical' entity type

Techniques for Refining the Model

(Cover Level, No Claims Bonus, etc) and a separate entity type created to reflect not only the current situation but the full history as well. This is shown in Figure 4.24. This includes a COVER entity type and things are getting better. But this is not the end of the story. The effect of this additional entity type on the existing relationships in which POLICY participates should be considered to see if any should now be moved. In this case, adjustment is required and Figure 4.25 shows the result. It is fairly self-evident that payments will become due on a POLICY normally when there is some change in the details relating to COVER, even if this is merely a straightforward renewal. The relationship needs to be moved to ensure that it ties the data together correctly. By leaving PAYMENT DUE related to POLICY there could well be difficulty relating payments to COVER changes, which would be less than satisfactory.

This has removed the need to update the data. When there is a change a new COVER record can be added. Information cannot be lost and a full history has been preserved. The separation of the static from the volatile is also of benefit as the processes that act on each type are generally different. Indeed, the mixing of static and volatile attributes in one entity type is often indicative of inadequate data analysis and of the need to split the entity type concerned.

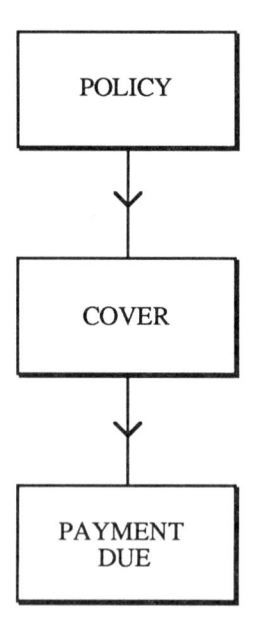

Figure 4.25   The effect of the historical entity type on existing relationships

Chapter 5

Checking the Model

Establishing Identifiers

All entity types must contain one or more attributes that can uniquely identify occurrences of the entity type. There may be:

* *one attribute that achieves this;*
* *several attributes in combinations that achieve this.*

If this cannot be done this can be resolved by:

* *checking for time stamps that have not been identified;*
* *conducting further analysis on attributes;*
* *'inventing' an attribute to promote discussion.*

We can now move on to a discussion of various techniques which can be used to ensure that the right attributes are in the right entity types. When attributes were discussed previously the concern was with Identifying, Supportive and Descriptive Attributes. This section concentrates purely on the first of these, namely Identifying Attributes.

Why is it important to establish Identifying Attributes? Not all database systems require that each occurrence of each entity type needs to be uniquely 'identifiable'. However, some do and there is value in developing the data model so that it is as independent as possible from the implemented solution. Also many systems will only allow direct access to occurrences of an entity type if individual occurrences can be uniquely identified. It is therefore good practice to ensure that each occurrence of each entity type can be identified uniquely.

So how do we set about ensuring that we have established the correct identifiers? In the majority of cases this is quite straightforward. Most of the major entity types will have fairly obvious identifiers. For example, customers will normally be identified by a Customer Number and this will be unique. The same will apply to Invoices, Parts and many other entity types. The question to ask is simple – "Of the attributes that have been established for this entity type, which attribute or combination of attributes will uniquely identify this occurrence of the entity type from all other occurrences?" As has been said, in many cases, one attribute will suffice to provide this uniqueness. But in others, it will be necessary to take a number of attributes in combination.

There is no problem with this, if necessary *all* the attributes can be used as identifiers. One thing to bear in mind here is the effect of time on the model. Often, where the entity type is time-dependent, the only thing that will provide this uniqueness are the dates applicable to the entity type. Consider the example used in the previous chapter (Figure 4.22). The COVER entity type will be identified by the policy number that it applies to. Therefore, there will be no immediate means of distinguishing between one COVER entity for a POLICY and any other. The uniqueness is provided by the start date in combination with the policy number. The combination of the two attributes will give the uniqueness required but only if the business is restricted to writing a maximum of one COVER per POLICY per day. This is probably acceptable but does indicate that care needs to be exercised over which attributes are used.

This brings up another point. If the identifiers are established without due regard then there is the danger of imposing constraints that are artificial. Sticking to the example, if it was required to retain the ability to write more than one COVER for a POLICY in a day, then not only is there duplication, but if the implemented database will not tolerate duplicates, and many will not, then it will be impossible to do it anyway.

Finally, there will be occasions where there do not seem to be sufficient attributes to allow the designation of one or more as identifiers and still have uniqueness. The usual reason for this is that the complete set of attributes has not yet been identified and further analysis is required. Again, it is normally true that such attributes are date or time-stamps, which may not have been thoroughly identified normally because the effect of time on the entity type has not been fully realised. On some rare occasions the data analyst will not be able to establish unique identifiers even after further analysis of the entity type. Here the data analyst has no option but to 'invent' a suitable attribute and add this into the entity type. This may sound odd, to suddenly 'make something up', but again it does draw attention to the area.

A completely new attribute is almost certain to promote discussion amongst users and may highlight the as-yet-undiscovered identifier. Care is needed though. This is a 'last resort' and should not be undertaken lightly; the data analyst should remember that this action has been taken and should return to the area to review the decision from time to time. But what is the nature of this invented attribute? Well, keep it simple and straightforward – a serial number or some code which does not carry much information content of its own. Anything too 'real' will tend to mask the fact that it has been so invented and make it harder to remove if and when the real identifier becomes apparent.

Supportive Attributes

> *Each relationship should be supported by propagating the Identifying Attributes of the entity type at the 'One' end of the relationship, into the entity type at the 'Many' end. Normally this is straightforward, but note the following:*
>
> • *the same attribute may get propagated several times into the entity type. In this situation the actions are:*
>   – *if the values will be the same, show the attribute only once*
>   – *if the values may differ, consider the use of a 'Role' entity type;*
> • *the model may become unwieldy due to multiple propagation. In these cases, insert an invented identifier, a 'propagation break', to solve the problem.*

When attributes were discussed in an earlier chapter the supportive role that they may play was considered. In effect, what is being said is that in any given entity type there must be appropriate attributes to support each relationship in which the entity type participates. The basis for this lies in the depths of relational theory and need not be discussed here but from a practical point of view there are still good reasons for ensuring this is done.

The data model is software-independent as it is being modelled from a business rather than an implementation standpoint. It must be ensured, though, that whatever form of database may eventually be chosen, the data model will be appropriate to that view. In many DBMS packages, especially relational databases, relationships are defined purely in terms of Supportive Attributes without recourse to pointers or other means of physical addressing. So we not only have the theory behind us but it makes good practical sense as well.

What action needs to be taken? It is necessary to consider each entity type in the data model and ensure that wherever it participates as the 'Many' in a One-to-Many or Zero-or-One-to-Many Relationship, then there are attributes that will identify the appropriate *occurrence* of the 'One' to which any *occurrence* of the 'Many' will relate.

The diagram shown as Figure 5.1 will make this clearer. MAKE has an identifying attribute, Make Code, and any number of descriptive attributes not shown. MODEL not only has Model Code as a part identifier, but assuming here that this is non-unique, the Make Code is also needed as an identifier. The two attributes together achieve the uniqueness required but the inclusion of the Make Code also supports the relationship to MAKE. For any occurrence of MODEL, there is an attribute that identifies the occurrence of MAKE to which it is related.

```
        ┌────────┐        ┌────────┐
        │  MAKE  │        │ ENGINE │
        │        │        │  TYPE  │
        └────┬───┘        └────┬───┘
             │                 │
             ▼                 │
        ┌────────┐             │
        │ MODEL  │             │
        └────┬───┘             │
             │                 │
             ▼                 │
        ┌────────┐◄────────────┘
        │ MODEL  │
        │  HAS   │
        │ ENGINE │
        └────┬───┘
             │
             ▼
        ┌────────┐
        │  CAR   │
        └────────┘
```

MAKE   [Make Code, Make Name, etc]

MODEL   [Make Code Model Code, Model Name etc]

ENGINE TYPE   Engine Code, Size, Fuel type etc]

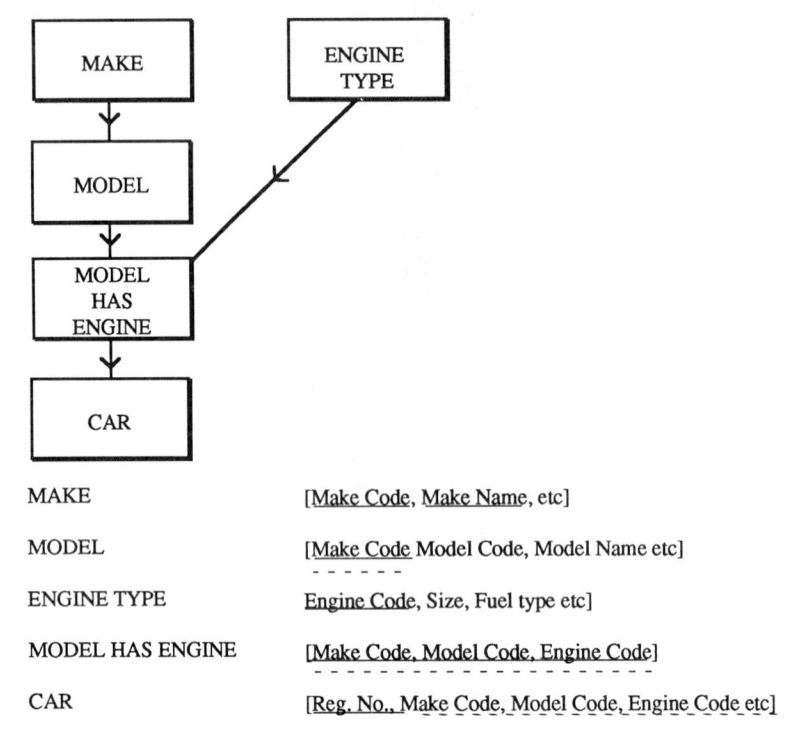

MODEL HAS ENGINE   [Make Code, Model Code, Engine Code]

CAR   [Reg. No., Make Code, Model Code, Engine Code etc]

Figure 5.1   A small model showing supportive attributes

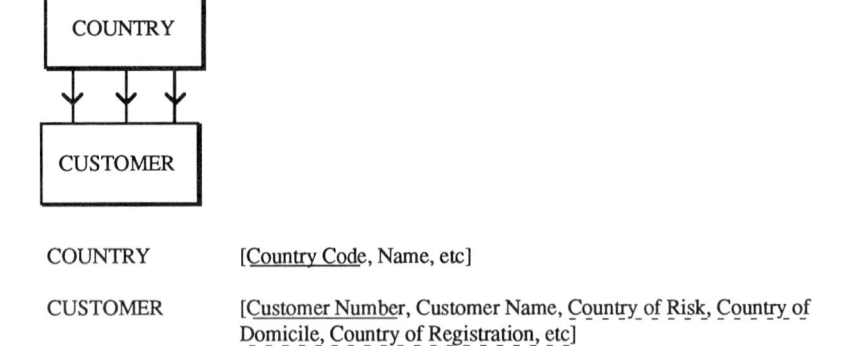

```
        ┌──────────┐
        │ COUNTRY  │
        └─┬──┬──┬──┘
          │  │  │
          ▼  ▼  ▼
        ┌──────────┐
        │ CUSTOMER │
        └──────────┘
```

COUNTRY   [Country Code, Name, etc]

CUSTOMER   [Customer Number, Customer Name, Country of Risk, Country of
               Domicile, Country of Registration, etc]

Figure 5.2   Multiple occurences of an attribute with different values

ENGINE TYPE is identified by Engine Code and has other purely descriptive attributes. The next entity type, MODEL HAS ENGINE, is a cross-reference entity type and requires the identifiers from both MODEL and ENGINE TYPE as its own identifiers. Again, these attributes are also acting as supportive, the Make Code and Model Code identify the MODEL and the Engine Code identifies the ENGINE TYPE. The last entity type, CAR, is slightly different. Here, uniqueness is conferred by the single attribute Registration Number, but this does not support the relationship in which CAR participates. It is necessary, therefore, to propagate the Identifying Attributes from MODEL HAS ENGINE into CAR in order to support the relation. These are not Identifying Attributes for CAR, but they are Supportive. This propagation must be carried out on each entity type in the model in order to be thorough in approach.

There are, however, two particular problems that can arise in this exercise. Firstly, this can often result in a complex model, with the same attribute apparently appearing a number of times within an entity type. Where this happens it must be asked if the values of the attributes will be the same or different for any given occurrence of the entity type. Figure 5.2 shows an example where they may be different, Figure 5.3 is an example where they would be the same. In Figure 5.2 there are three relationships between COUNTRY and CUSTOMER and each is for a different purpose and could conceivably have different values. In this type of situation, each relationship must be separately supported and three values for the attribute are required. This can be simplified as shown in Figure 5.4. By making use of one of the earlier techniques, the Role entity, the need to support all three relationships in CUSTOMER has been removed. Further, should a new role be added, no changes to the attributes of any of the entity types is required.

Figure 5.3 on the other hand, is a more complex situation altogether. Here there are a number of options available for each ACCOUNT. The choice of options is given in the entity types INTEREST STATEMENT FREQUENCY, ACCOUNT STATEMENT FREQUENCY and INTEREST POSTING FREQUENCY. The applicable values that these may take are independent of each other but are restricted by the ACCOUNT TYPE, and therefore the ACCOUNT must relatet to all three cross-reference entity types. If this is so, then the identifiers of each need to be propagated into ACCOUNT as Supportive Attributes. Here there is no need for three occurrences of Account Type Code in ACCOUNT as logic dictates that the three values for Account Type Code must be the same. The attribute should, therefore, be shown only once.

However, a glance through the attributes here will show that this data model can be resolved into a cleaner structure. The resolution is shown in Figure 5.5. Here the three options have been combined into one entity type and the Frequency Code has been split out so that by cross-referencing these

```
                                    ACCOUNT
           INTEREST                 ACCOUNT         ACCOUNT         INTEREST
           STATEMENT                STATEMENT       TYPE            POSTING
           FREQUENCY                FREQUENCY                       FREQUENCY


           INTEREST                 ACCOUNT                         POSTING
           STATEMENT                STATEMENT                       FREQUENCY
           FREQUENCY                FREQUENCY                       FOR ACCOUNT
           FOR ACCOUNT              FOR ACCOUNT                     TYPE
           TYPE                     TYPE


                                            ACCOUNT
```

INTEREST STATEMENT FREQUENCY   [Frequency Code, Description etc]

ACCOUNT STATEMENT FREQUENCY   [Frequency Code, Description, etc]

ACCOUNT TYPE   [Account Type Code, Description etc]

INTEREST POSTING FREQUENCY   [Frequency Code, Description, etc]

INTEREST STATEMENT FREQUENCY   [Frequency Code, Account Type Code]
FOR ACCOUNT TYPE

ACCOUNT STATEMENT FREQUENCY   [Frequency code, Account Type Code]
FOR ACCOUNT TYPE

POSTING FREQUENCY FOR   [Frequency Code, Account Type Code]
ACCOUNT TYPE

ACCOUNT   [Account Number, Interest Statement
                                          Frequency Code, Account Statement Frequency
                                          Code, Posting Frequency Code, Account Type
                                          Code etc]

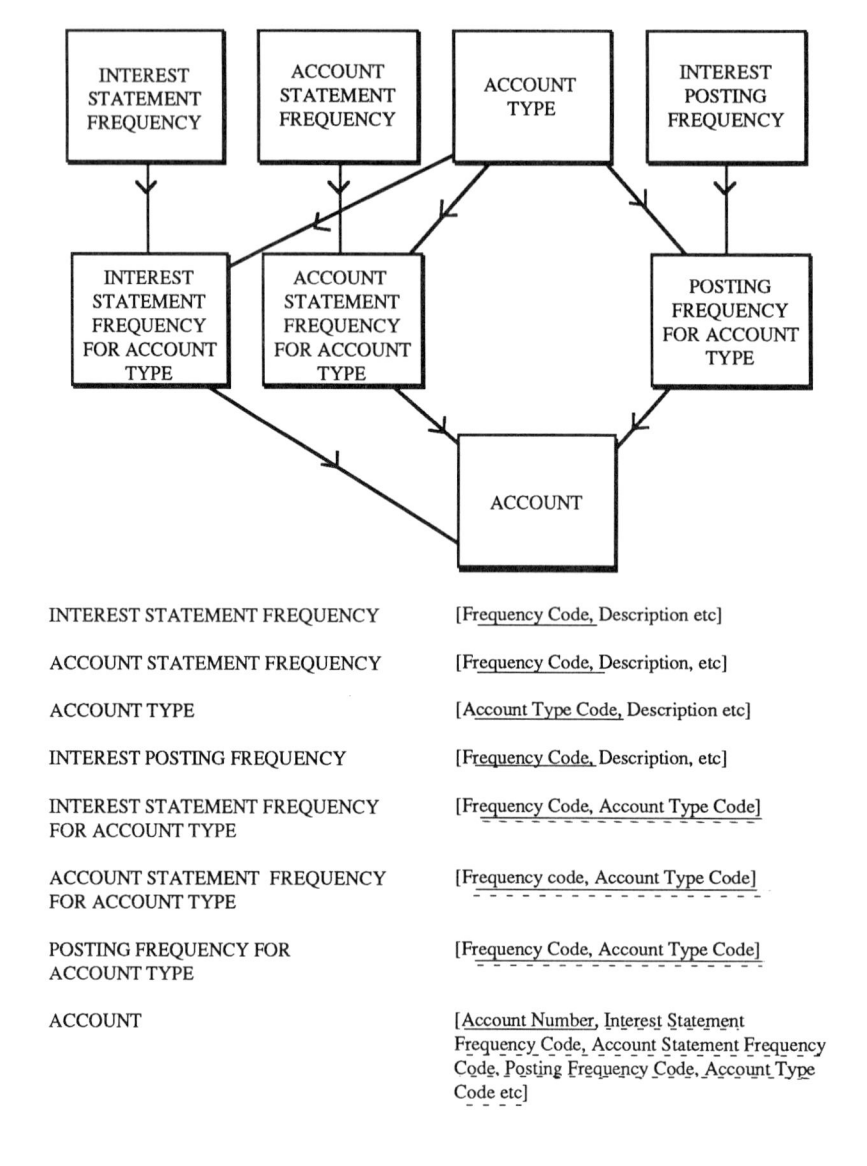

Figure 5.3   Multiple propagation where the values are the same

two with ACCOUNT TYPE the range of PERIODS for each PERIODIC FUNCTION that may apply to ACCOUNTS of a given ACCOUNT TYPE can be defined. The actual PERIODIC FUNCTION and PERIOD for a specific ACCOUNT is then cross-referenced below this. This gives a cleaner structure but also problems. There is nothing in the model to prevent an ACCOUNT relating back through the cross-references to the same PERIODIC FUNCTION for more than one PERIOD. This is not so much a problem with the Data Structure but more to do with the inadequacy of the modelling convention inasmuch as there is no means of reflecting this constraint. All that can be done is note that the constraint exists and ensure that it is *programmed* in at implementation time. It is an imperfect world!

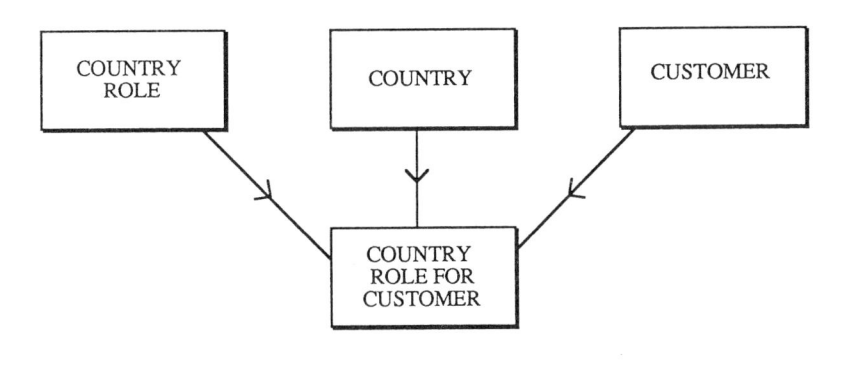

COUNTRY ROLE	[Role Code, Description etc]
COUNTRY	[Country Code, Name etc]
CUSTOMER	[Customer Number, Name etc]
COUNTRY ROLE FOR CUSTOMER	[Role Code, Customer Number, Country Code]

Figure 5.4   Resolution of Figure 5.2

However, back to the problems associated with attribute propagation. The second problem that occurs is that in some situations the list of attributes being propagated grows and grows as one moves down through the structure. This is invariably caused by a lack of Identifying Attributes native to an entity type, resulting in each entity type having the Identifying Attributes of the entity types above, it propagated into it and then further propagated into the entity types below. Where this occurs it is necessary to introduce an artificial 'propagation break'. This is done by inventing an identifier that will provide uniqueness for an entity type at a suitable point in the structure. Following the guidelines in the previous section, this will cause no problems and prevent the endless propagation of attributes as only this identifier need be included in the next level down. There is no hard and fast rule for deciding when this is necessary, but it is normally quite easy to detect when the entity types are becoming too full of propagated attributes that have little relevance to the entity type under consideration.

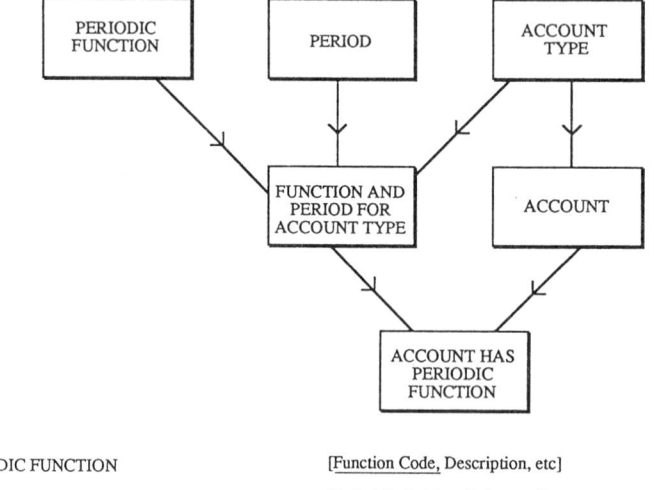

PERIODIC FUNCTION	[Function Code, Description, etc]
PERIOD	[Period Code, Description, etc]
ACCOUNT TYPE	[Account Type Code, Other attributes]
FUNCTION AND PERIOD FOR ACCOUNT TYPE	[Period Code, Function Code, Account Type Code]
ACCOUNT	[Account Number, Account Type Code, other Attributes]
ACCOUNT HAS PERIODIC FUNCTION	[Period Code, Function Code Account Type Code, Account Number]

Figure 5.5 Resolution of Figure 5.3

Alternate Keys

It is necessary to check that sufficient attributes are included to support any required access paths. No other action needs to be taken. Such keys do not affect the choice of Identifying or Supporting attributes.

This section is included not so much because it is a *bona fide* data analysis technique but rather because many people often ask about determining alternative keys when learning about data analysis.

So what is meant by 'Alternate Keys'? The identifying attributes provide a unique primary 'key' for the entity type. If the values of the identifiers are known the appropriate entity occurrence can be selected. Alternate keys, then, are any other attributes or combination of attributes that may be used to access occurrences of the entity type and they may or may not be unique. For example, if it is required to hold Customer-related detail there will almost certainly be a CUSTOMER entity type, with a unique identifier such as Customer Number. Although this is unique, it is not particularly helpful as a search key and an abbreviated name or mnemonic may be needed to facilitate selection. This secondary key would not necessarily be unique and can be treated as any other attribute. It becomes important only at the design stage when access has to be possible via this key.

Although this is often seen as a problem, it is virtually irrelevant to data analysis. The data model requires that there is a unique identifier for each occurrence of the entity type and the first or primary key will do this. Any other 'keys' are merely additional attributes which will need to be included but it is not necessary to take any specific action over them.

The requirement for these Alternate Keys will emerge from Process Analysis, a topic that will be examined later in this book, and to a certain extent is implementation-driven. Only if there is a choice of unique identifiers does this need to be considered. Where there is such a choice, the attribute, or set of attributes used as identifiers should be chosen on the grounds of which is the most straightforward set to choose. The decision is to some extent arbitrary and will not affect the decision about providing access points which will be taken during implementation design.

Suffice it to say that, if there is a recognised need for an Alternate Key, then the appropriate attributes must be included in the entity type during the data analysis phase. But it is by no means essential that these attributes be any more than ordinary Descriptive Attributes of that Entity type.

Normalisation

Normalisation asks three questions in order to establish an 'ideal' data structure. These will:

* *remove repeating attributes;*
* *ensure that all parts of the Identifier are needed;*
* *remove interdependencies between attributes.*

The penultimate section in this Chapter turns attention to a technique often associated with data analysis. The technique is known as Normalisation and consists of a number of tests which can be applied to the attributes of an entity type. The results of these tests will then lead to an 'ideal' allocation of attributes and entity types. It is fair to say that there has been much theoretical argument and debate concerning the validity of Normalisation and those who are interested will be able to find considerable discussions elsewhere. It is the purpose of this book to concentrate on practical rather than theoretical issues and Normalisation makes explicit some of the rules and tests that need to be applied during the data analysis process.

With Normalisation three precise questions are asked about each attribute. Application of the results leads to the data being in First, Second and finally Third Normal Form. So what are the questions?

1 Does the attribute occur several times within the entity type?

2 For entity types with multi-attribute identifiers only, is it possible to establish the *value* of the attribute, if the *values* of only some of the identifying attributes are known?

3 Is it possible to establish the *value* of the attribute if the *value* of any other Descriptive or Supportive attribute is known?

If the answer to all three questions is "No," then that entity type is fully normalised, or in Third Normal Form [3NF].

These questions are examined in more detail below and are worked through using the data model shown in Figure 5.6.

First Normal Form [1 NF]

Does the attribute occur several times within the entity type? It can be seen that both Addresses for CUSTOMERS and Order Items within the ORDER are repeating groups and as such violate First Normal Form. If no action is taken the problems outlined in the section covering Entity Splitting in Chapter 4 will occur. It is not known how many occurrences of the group have to be allowed for; if a restriction is placed on this, then a fairly major constraint has

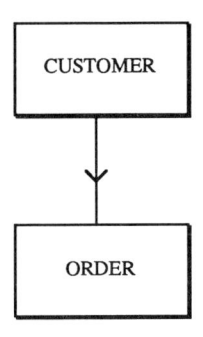

```
                    ┌──────────┐
                    │ CUSTOMER │
                    └────┬─────┘
                         │
                         ▼
                    ┌──────────┐
                    │  ORDER   │
                    └──────────┘
```

CUSTOMER   [Customer Number, Customer Name, Salesman Id, Salesman Name, Address 1, Address 2, ... Address n, etc]

ORDER   [Order No., Customer No. Date, Part No 1, Quantity 1, Description 1, Part No. 2, Quantity 2, Description 2, ... Part No. n, Quantity n, Description n, etc]

Figure 5.6   An unnormalised Data Structure

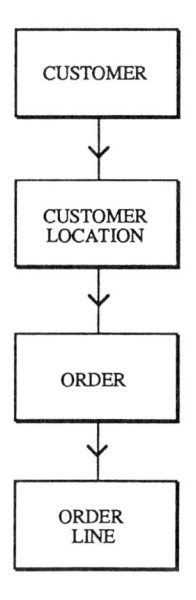

```
                    ┌──────────┐
                    │ CUSTOMER │
                    └────┬─────┘
                         │
                         ▼
                    ┌──────────┐
                    │ CUSTOMER │
                    │ LOCATION │
                    └────┬─────┘
                         │
                         ▼
                    ┌──────────┐
                    │  ORDER   │
                    └────┬─────┘
                         │
                         ▼
                    ┌──────────┐
                    │  ORDER   │
                    │   LINE   │
                    └──────────┘
```

CUSTOMER   [Customer Number, Customer Name, Salesman Id, Salesman Name, etc]
CUSTOMER LOCATION   [Customer Number, Location No., Address, etc]
ORDER   [Order No, Customer No, Location No, Date, etc]
ORDER LINE   [Order No, Part No, Quantity, Description]

Figure 5.7   A data structure in First Normal Form [1NF]

Checking the Model

been imposed on the eventual implementation. The solution is to break out the repeating group to a separate entity type related to the original one and reconstruct the model. This is shown in Figure 5.7.

Second Normal Form [2NF]

For entity types with multi-attribute identifiers, is it possible to establish the *value* of the attribute if the *values* of only some of the Identifying Attributes are known? This test will only apply to the entity types, CUSTOMER LOCATION and ORDER LINE. CUSTOMER LOCATION passes the test, the values of both Customer Number and Location Number must be known in order to establish the Address. The entity type ORDER LINE does not pass. If the value of Part Number is known, the value of Description is also known. If this is not corrected, the Description will be repeated in every ORDER LINE occurrence for each ORDER that features that part. This will lead to considerable redundancy and leaves the door open for various forms of inconsistency to creep in. There is no control over what value the Description can take. To solve this the attribute and its identifier should be broken out into a separate entity type and again the model needs to be reconstructed. This is shown in Figure 5.8.

Third Normal Form [3NF]

Is it possible to determine the *value* of the attribute if the *value* of any other Descriptive or Supportive attribute is known?

All the entity types pass this test with the exception of CUSTOMER. Given the value of Salesman Id, the value of Salesman Name is defined, that is Salesman Id defines Salesman Name. If this is not corrected there will again be the danger of inconsistent information and redundant data. Salesman Name will be repeated in every CUSTOMER with which the Salesman deals. Again the solution is to split the offending attributes out into a new entity type. When this has been done, the model is said to be in Third Normal Form and the complete structure is shown in Figure 5.9.

Normalisation is often proposed as a formal method which must be applied as an essential step in data analysis. However, many of the techniques outlined in this book will lead to the creation of a fully normalised data structure in a more informal manner. Others would argue that the techniques explicitly mentioned have been used intuitively by good systems analysts for years anyway. In this approach the value of Normalisation lies in its use as a cross-check towards the end of the data analysis phase. It is especially useful

```
                    ┌──────────┐
                    │ CUSTOMER │
                    └────┬─────┘
                         ▼
                    ┌──────────┐
                    │ CUSTOMER │
                    │ LOCATION │
                    └────┬─────┘
                         ▼
                    ┌──────────┐       ┌──────┐
                    │  ORDER   │       │ PART │
                    └────┬─────┘       └───┬──┘
                         ▼                 │
                    ┌──────────┐◄──────────┘
                    │  ORDER   │
                    │  LINE    │
                    └──────────┘
```

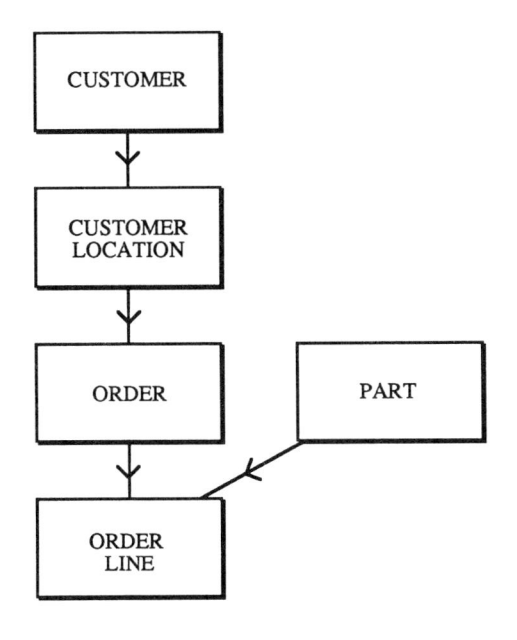

CUSTOMER   [Customer Number, Customer Name, Salesman Id, Salesman Name, etc]

CUSTOMER LOCATION   [Customer Number, Location Number, Address, etc]

ORDER   [Order Number, Customer Number, Location Number, Date, etc]

PART   [Part Number, Description, etc]

ORDER LINE   [Order Number, Part Number, Quantity, etc]

Figure 5.8   A Data Structure in Second Normal Form [2NF]

where the major information source has been an existing computer system as traditional DP record layouts are full of unnormalised data. It is not justifiable to dismiss Normalisation but neither should its value be over-stressed. The best approach is to apply the rules explicitly until they become second nature at which point you will find little return from the effort of using them mechanically.

To complete this section, just a few comments on further types of Normalisation. Some researchers have suggested other developments from Normalisation theory. (Including Fourth Normal Form [4NF] and Boyce-Codd Normal Form [BCNF].) With a thorough application of the techniques within this book and a clear understanding of the underlying principles, these further developments will be unlikely to be of benefit.

However, data analysis is still a developing art both in theoretical and practical terms and the data analyst must at all times remain receptive to new ideas and revised techniques.

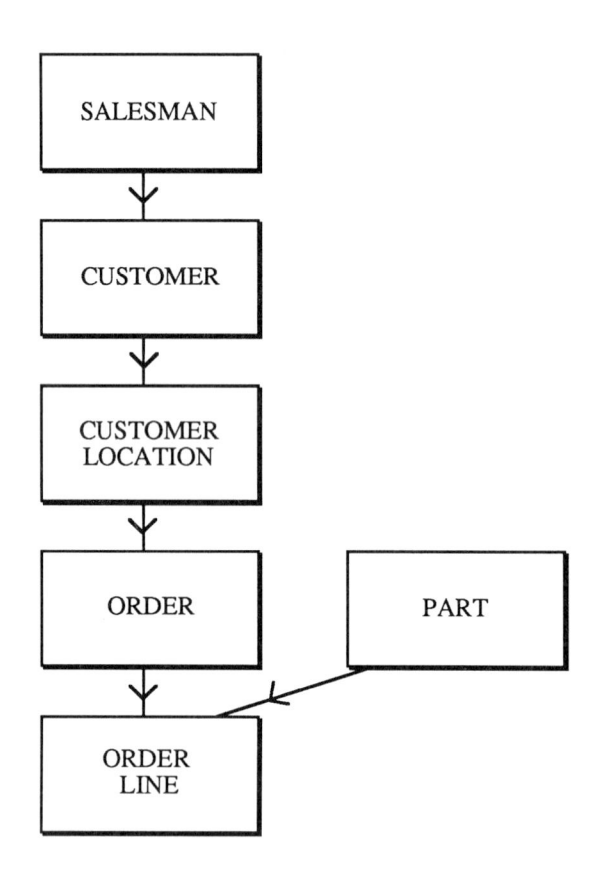

```
            ┌─────────────┐
            │  SALESMAN   │
            └─────────────┘
                   │
                   ▼
            ┌─────────────┐
            │  CUSTOMER   │
            └─────────────┘
                   │
                   ▼
            ┌─────────────┐
            │  CUSTOMER   │
            │  LOCATION   │
            └─────────────┘
                   │
                   ▼
            ┌─────────────┐        ┌─────────────┐
            │   ORDER     │        │    PART     │
            └─────────────┘        └─────────────┘
                   │                      ╱
                   ▼                     ╱
            ┌─────────────┐◄────────────
            │   ORDER     │
            │   LINE      │
            └─────────────┘
```

SALESMAN	[Salesman Id, Salesman Name, etc]
CUSTOMER	[Customer Number, Customer Name, Salesman Id, etc]
CUSTOMER LOCATION	[Customer Number, Location Number, Address, etc]
ORDER	[Order Number, Customer Number, Location Number, Date, etc]
PART	[Part Number, Description, etc]
ORDER LINE	[Order Number, Part Number, Quantity)

Figure 5.9   A Data Structure in Third Normal Form [3NF]

Abstraction

The aim of Abstraction is to identify more powerful underlying concepts. This is done by searching for several entity types that can be 'collapsed' together, for example BUSES, CARS and LORRIES are all instances of VEHICLE. Dangers to bear in mind are:

* *combination of entity types will be difficult if the attributes differ greatly;*
* *differences in Abstraction level between different sections of the model. This can be identified by a large number of exclusive Zero-or-One-to-Many Relationships.*

The topic of the final section in this chapter, and also the last technique to be discussed, is Abstraction. There has already been some mention of this technique in the discussion on the modelling of hierarchies in Chapter 3.

The technique of Abstraction is probably the most powerful single technique that can be used. Unfortunately, it is also extremely difficult to explain in concise terms. The aim is to define the entity types within the model so that they are intrinsically flexible. This is usually achieved by determining what linguists call 'deep meaning' and this will be described by means of a simple example. The Insurance model may initially have started with something similar to Figure 5.10. This model allows the recording of details of BUSES, CARS and VANS and their terms of insurance. But this model is intrinsically weak. If it is required to restrict the mix of insured risks on a POLICY then there is no way of achieving this; secondly, it is a cluttered structure; thirdly, if a new type of vehicle is to be covered, LORRY, the model has to change to accommodate it.

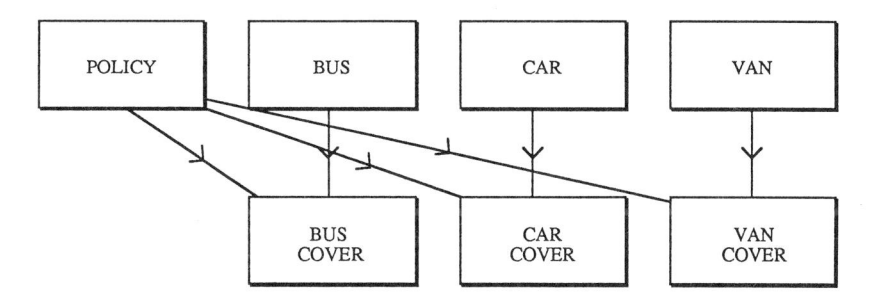

Figure 5.10 A Data Model with 'shallow' entity types

Checking the Model

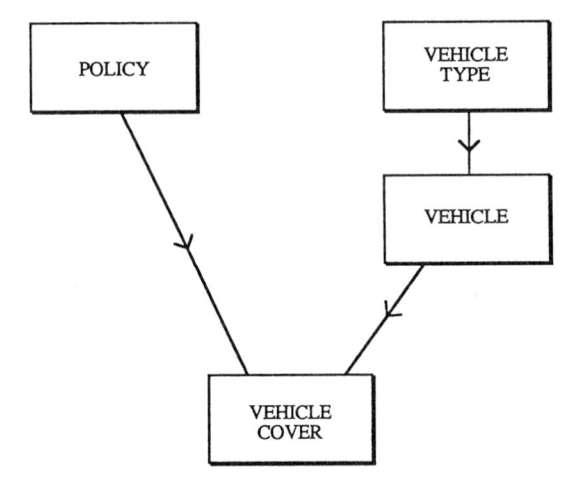

Figure 5.11 A Data Model with 'deep meaning'

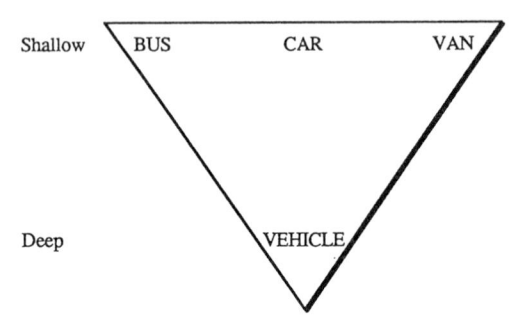

Figure 5.12 The 'shallow' concepts are instances of the 'deeper' concepts

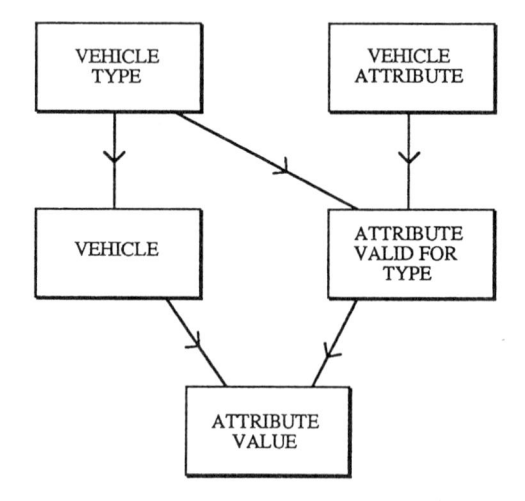

Figure 5.13 A mechanism for dealing with variable attributes

The skill is in recognising that BUSES, CARS and VANS are all instances of the same, more fundamental, concept: VEHICLE. Once this is realised it is possible to construct a much more resilient model as shown in Figure 5.11. Here the concepts which were entity types in Figure 5.10 (CAR, BUS, VAN) can be expressed as *entity occurrences* at the VEHICLE TYPE level, and one concept, VEHICLE, serves the purpose of representing occurrences of all cars, vans and buses. This is much more flexible. To insure lorries, a new VEHICLE TYPE occurrence can be added, and the model is then able to handle this new concept.

In terms of 'deep' and 'shallow' meaning this can be represented as an inverted triangle as in Figure 5.12. As can be seen there are many fewer concepts at the lower level. These concepts are both more powerful and more fundamental. If a model can be constructed around such concepts the structure will be cleaner, more elegant, and far more flexible.

But what are the dangers? First of all, the technique can give problems with attributes. The approach only works easily if there is fair commonality between the attributes of the 'shallow' concepts. Where there is a great variety of attributes the situation is more difficult. Assume for the moment that the attributes of buses, vans and cars are very different. By combining them all into VEHICLE there will be many optional attributes and it can be difficult to restrict their usage to the appropriate VEHICLE TYPES. It can be done though, as shown in Figure 5.13. This can be modelled satisfactorily and is still flexible, but generating programs to manipulate this structure can be problematic, so beware.

Secondly, this technique can cause problems with relationships. Where this occurs it is usually due to a difference in abstraction levels between different parts of the model. To demonstrate this, the Motor Insurance Model has been 'un-abstracted' *(see Figure 5.14)*. This has regressed back to CARS, BUSES and VANS and also POLICY has been 'un-abstracted' into three different entity types. If it is realised that it is possible to abstract on VEHICLE but it is not noticed that the same can be done to POLICY immediate problems arise as Figure 5.15 makes clear. There are now three exclusive Zero-or-One-to-Many Relationships and the structure is awkward. By understanding the problem of abstraction levels we can abstract on POLICY and return to the elegant structure we had in Figure 5.12.

This brings up a very important point. It is quite likely, in a real situation, that BUS, CAR and VAN insurance would be carried out by different parts of the company. This could easily lead to the development of a model which had different sections (Figure 5.14) as each part of the company would see things differently. If this happened and if the company did decide to move into lorry insurance there would be a whole new section to model. So, in practice, abstraction is not simply about collapsing similar entity types together, this is relatively easy; rather the trick is to collapse whole sections

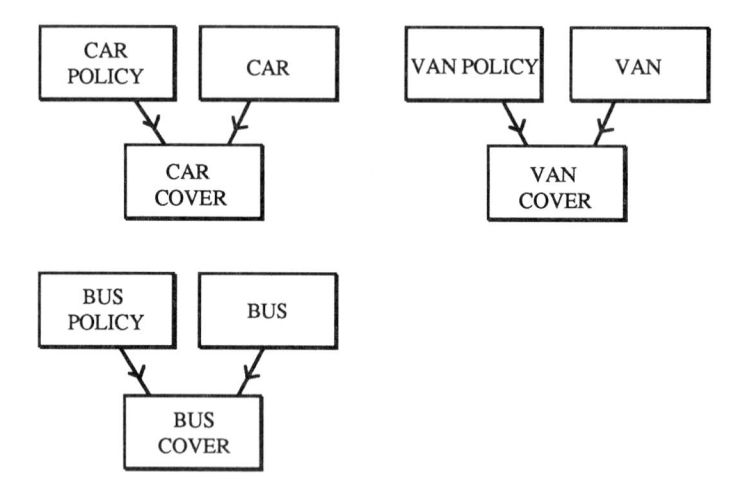

Figure 5.14   A very 'shallow' model

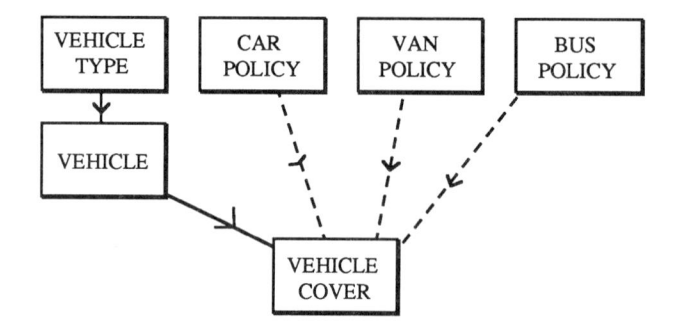

Figure 5.15   A Data Model with variable abstraction level

of the model if these have been built on 'shallow' concepts. This is far more difficult and requires practice. Wherever possible, these more powerful concepts should be those that appear in the model, but remember, the user may not employ these concepts when talking to the analyst .... they have to be dug out!

As a conclusion to this part of the book, we will finish with something very abstract! The ultimate in abstract data models. This model, shown in Figure 5.16, will handle *any* application .... at least in theory! It illustrates a number of points about abstraction: first of all, an abstract model is harder to understand; secondly, it is extremely flexible; thirdly, it is difficult to write software to handle it! If you have understood the concepts and techniques discussed so far, this model should make sense although it needs some thought. Any entity type can be described as an occurrence of ENTITY TYPE (for example Customer, Account, Policy). All occurrences of every entity type are identified as occurrences of ENTITY OCCURRENCE. The OPTIONAL/MANDATORY entity type serves two purposes. At the attribute level it states if an attribute must or may not have a value for an occurrence of an entity type. At the relationship level it distinguishes between the two types of relationship, One-to-Many (Mandatory), and Zero-or-One-to-Many (Optional). All possible attributes are occurrences of ATTRIBUTE and are cross-referenced to the ENTITY TYPES that use them.

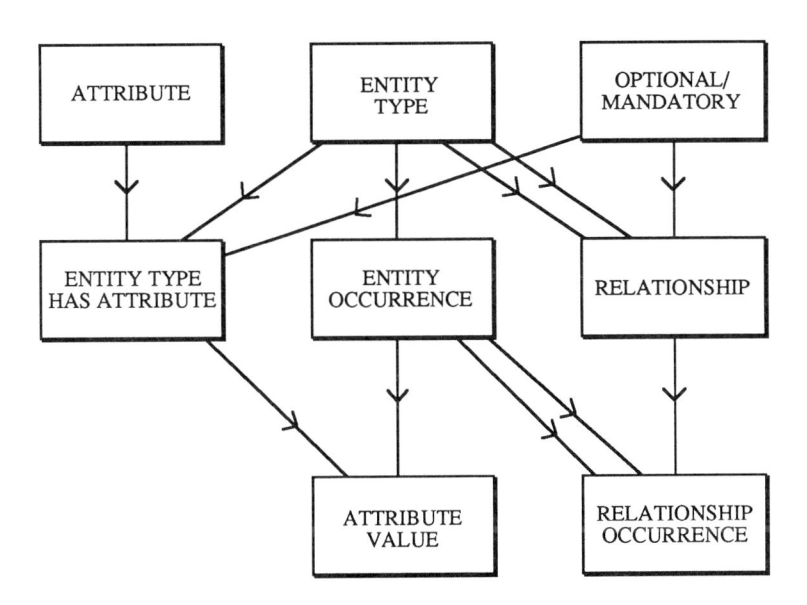

Figure 5.16 The 'Ultimate' Data Model

Checking the Model

The value taken by an ATTRIBUTE for a specific ENTITY OCCURRENCE is held in ATTRIBUTE VALUE. Finally, relationships are defined in terms of the ENTITY TYPES they link and related entity occurrences are identified in RELATIONSHIP OCCURRENCE. This model is not complete. It can be extended to deal with attribute roles and various other factors, but it does illustrate how far we can go with this extremely powerful technique.

Part 2

Using Data Analysis

Chapter 6

The Data Model

Diagram Conventions

Certain 'rules' apply to model presentation. These are:
- *omit date as an entity type, but clearly name date dependent attributes;*
- *draw relationships in a 'top-down' manner;*
- *do not worry about naming relationships.*

We have now completed our discussion of data analysis techniques and you will have become familiar with the conventions for representing entity types and relationships but, so far, the techniques have been examined in pure isolation. How these techniques are used in order to successfully carry out a data modelling exercise has not been considered. This part of the book examines a number of issues concerned with actually *using* the techniques. This covers how to present the model, developing ideas, making use of existing systems, involving users and coping with large complex models.

First, let us consider presenting the model and in particular further conventions for drawing it. In all the examples so far, the number of entity types shown has been very small, five or six maybe. Even when modelling a small discrete area, the number of entity types is likely to be far higher; say from fifteen to several hundred. In these situations it is useful to have some standard way of presenting the model to aid understanding.

Although all relevant entity types would normally be shown on the data model diagram, there is one very important exception. The importance of time on the model has been stressed in the earlier chapters and it is usual for there to be many date-dependent entity types and relationships in any model. This can give many problems. The number of lines that have to be drawn from DATE to other entity types can be very confusing and lead to a very cluttered appearance. Also there are likely to be entity types that have four, five or more relationships to DATE, another source of confusion and complexity.

The easiest way to cope with this is to simply miss DATE out of the model altogether. This may seem extreme, but there are so many occasions where DATE is used that it can almost be taken for granted. Each date attribute will have to be shown in the back-up documentation anyway and it is reasonable to assume that wherever there is a date attribute then there is a relationship between that entity type and an invisible DATE entity. There is one additional complication here. When naming date attributes one should take care to ensure that the name chosen implies the nature of the relationship: don't use Date 1, Date 2 ..... but rather Inception Date, Maturity Date, Review Date, and so on. This really applies to any attribute, but it is especially important in this instance.

Secondly, take care with the direction of relationships. Many data modelling conventions allow entity types to be placed anywhere on the page that is most convenient. The result of this is that relationships run from the 'one' to the 'many' in all directions, up, down, left to right, right to left, etc. This means that the reader has to concentrate on the line itself in order to establish which entity type is the 'one' and which the 'many'. It is less convenient but far clearer if relationships are always arranged so that they point downwards from the 'one' to the 'many'. This allows a clear 'top-down' approach to the model with the categorising and codifying entity types towards the top of the page, and the actual data containing entity types at the bottom. Don't worry about crossing lines. It's quite permissible and usually not confusing. The aim is to show the structure and relationships as clearly as possible, not to worry about artistic niceties.

There is a further benefit of this downward relationship approach. This book has discussed the importance not only of studying the data but also the rules that the data must obey. When these are drawn out, it will be found that the fairly static rules will be shown at the top and the more volatile data at the bottom. There will be a gradually increasing degree of volatility inherent in the entity types moving down the structure. This again functions as an aid to comprehension and gives a more natural feel to the model than would be the case if the entity types were placed anywhere on the paper.

The last point in this section concerns relationship names. Many conventions allow or insist that all relationships be named, normally with a suitable verb to give meaning to the relationship. For example MAN 'drives' CAR, where 'drives' is the relationship. This causes no problems on a simple model and has some use, but as the model starts to develop it can be tiresome and confusing to name each relationship individually. In most cases, the relationships are quite self-explanatory anyway and where they are not, the back-up documentation, which is covered in the next chapter, can make this clear. A complex model is often much clearer without named relationships, rather than having the name written alongside the line, often on an angle, and generally adding clutter to the structure diagram. But it is not a hard and fast rule. Try it and see which you prefer!

Spotting Problems

Check for the following situations:

- *central entity types with many relationships: can the relationships be moved? Are they in the right place?*
- *multiple direct paths between two entity types: consider the use of 'Role' entity types;*
- *multiple indirect paths: are they all required? Do they lead to inconsistencies?*
- *Zero-or-One-to-Many Relationships: should the entity type be split to resolve this?*
- *similar Relationships: if two or more entity types participate in the same relationships, combination should be considered.*

Having drawn out the data model there are a number of things to look for which will indicate possible problems with the data analysis. On some occasions the Model will simply 'feel' wrong; although it can be difficult logically to pinpoint such problems they should not be ignored but investigated further to resolve the difficulty. However, a number of indicators can be pinpointed for further examination and a discussion of these follows.

Firstly, with almost all data models there will be one or two crucial and central entity types to which most other entity types will appear to be related. The usual manifestation of this is that there will be many relationships coming into, and out of, these central entity types. The example shown as Figure 6.1 is typical of the sort of situation that arises. Nearly every entity type on the model is related to the central entity type of CAR.

This must ring warning bells and the model should be examined to see if a greater degree of elegance can be introduced. The aim here is to 'explode' the relationships, to move them either higher up the structure or lower down. In general, ask the question "Does this relationship belong here, or would it be better moved elsewhere?" The result of this on the current example is shown in Figure 6.2. The number of relationships in which CAR participates has been considerably reduced and some fairly useful entity types that emerged from the process of rationalisation have been added. This model as shown is not without its shortcomings (what if a GARAGE has no franchise or the policy holder is not the vehicle owner?) but it is now a generally more useful and flexible structure.

It is true to say that thorough application of the techniques in Chapters 3-5 would lead to a structure as in Figure 6.2. By studying the diagram it is possible to cross-check that the techniques have been applied correctly.

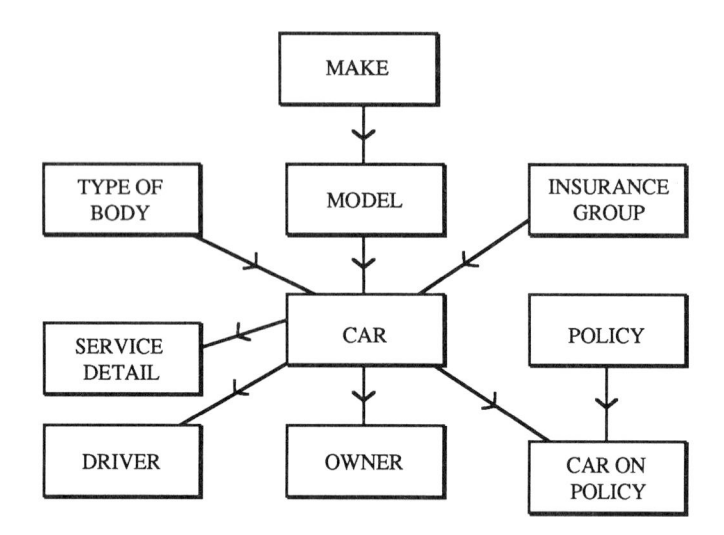

Figure 6.1 A Central Entity Type with many relationships

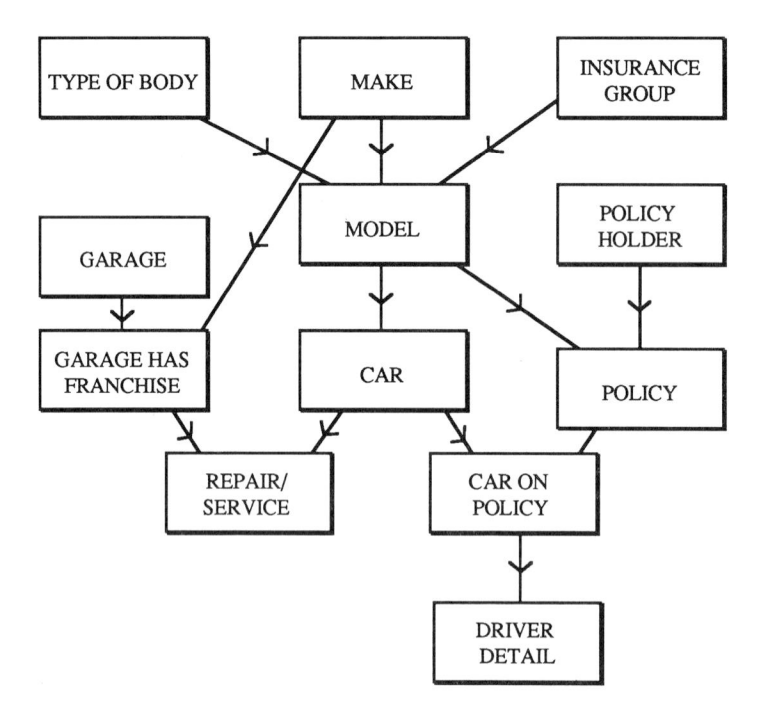

Figure 6.2 A more 'correct' version of Figure 6.1

The next problem area consists of multiple paths between the same two entity types. These fall broadly into two categories which may also be intermixed. The difficulty is that where there is a choice of routes between the two entity types it is necessary to ask that, for a given occurrence of 'A', is it required to get the same or different occurrence or set of occurrences of 'B'? The answer to this will affect the type of resolution. The first category is direct relationships, that is there is more than one relationship between two entity types with no intervening entity types. This situation has been discussed in Chapter 4 in the section on 'Role' entities and it is not proposed to discuss it further here.

The other category concerns indirect relationships. Here there are several routes from 'A' to 'B' but via other intervening entity types. The same questions need to be asked however, but the solutions may not be so clear cut. Can the relationships be combined in some way? Are they in the right place? It is not wrong to leave multiple paths in the model but they should be questioned. An example of this latter situation can be seen in Figure 6.2., where there are two routes from REPAIR/SERVICE to MAKE. In most situations it would be expected that a service would be carried out at a GARAGE which is franchised for that MAKE of car and in these situations the related MAKE occurrence would be the same via both routes. But this is not inviolate. Emergency repairs, for example, could be carried out by any reputable GARAGE so this should probably be allowed to occur but any REPAIR/SERVICE that indicated such an inconsistency should be queried.

Next any Zero-or-One-to-Many Relationships in the model should be considered, since it *may* indicate that two different things have been combined together in one entity type. Try splitting the entity type to create One-to-Many Relationships and examine the effects that this has on the rest of the model. If this creates problems elsewhere or makes the model unduly cumbersome, then it is correct to leave the Zero-or-One-to-Many Relationship untouched.

Lastly, look for two or more entity types that participate in the same or similar relationships. If such situations can be found it is *likely*, but not certain, that the entity types could be combined leading to a simplification of the model. But beware, if the relationships are not identical, combining the entity types could lead to some becoming Zero-or-One-to-Many Relationships as discussed above. In fact, there is often a trade-off between the two problems and both should be considered before a decision is made.

Numbering the Entity Types

In brief the rules to follow are:

- *start by selecting an entity type that depends on no others;*
- *number any entity types that depend only on those already numbered;*
- *select another non-dependent entity type;*
- *number any further dependent entity types, etc;*
- *attempt to number sets of related entity types together.*

In order to present the back-up documentation in a consistent way and in some intrinsic order, the entity types on the model should be numbered. This should take an essentially top-down approach, but at the same time it is necessary to pay some attention to any logical grouping of entity types that may exist. Start by selecting an entity type that does not depend on any other entity types, that is it does not participate as the 'many' in any relationship. Next, number any entity types that depend *only* on this entity type. Having done this, select another non-dependent entity type and assign the next number. Now number any entity types that depend only on entity types that have already been numbered and so on, gradually working down each 'leg' of the structure but only after all related higher level entity types have been numbered. This type of sequence makes it easier to change or revise bits of the model, to insert or remove entity types without disturbing the numbering sequence too greatly.

Figure 6.3. shows an example of how to number entity types. There are six potential start points to the sequence, CUSTOMER, INSURER, POLICY TYPE, ORG-UNIT, MAKE and VEHICLE TYPE. The most central entity type on the diagram is POLICY so it makes sense to work towards numbering this as early as possible. CUSTOMER is an obvious first choice followed by INSURER and POLICY TYPE. This enables INSURER ISSUES POLICY TYPE to be numbered thus allowing us to number POLICY. Now number COVER, as it depends only on POLICY, and ADDITIONAL DRIVER. Another start point must now be selected. Vehicle related information is obviously central and is probably more important (to the user) than ORG-UNIT. MODEL cannot be numbered until both MAKE and VEHICLE TYPE have been numbered, and as MAKE and MODEL would read better if presented contiguously, it is better to allocate the next number to VEHICLE TYPE, following on with MAKE, MODEL, VEHICLE and VEHICLE ON COVER. This leaves only ORG-UNIT and POLICY INVOLVEMENT which take the last two numbers. Easy isn't it?!

This takes a little practice to master and may seem like a lot of fuss over a small point, but it is necessary to take a consistent approach to presenting the documentation in order to maintain a coherent overall appearance. Mastering these conventions helps to impose an overall discipline on the documentation. The data model presented in the Appendix shows how these conventions are applied to a large model.

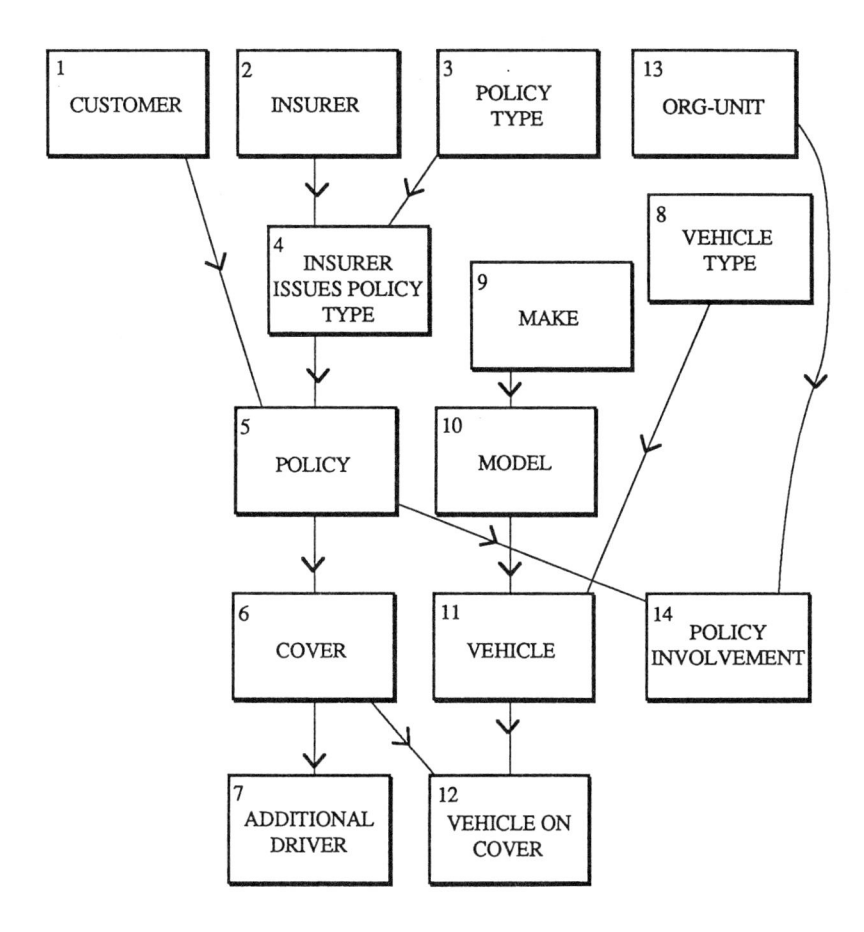

Figure 6.3   An example of entity type numbering

Quick and Dirty Models

It is permissible to break certain 'rules' in Quick and Dirty models:

- *only key entity types should be shown;*
- *Many-to-Many Relationships should be left unresolved;*
- *Recursive Relationships need not be broken down.*

Although useful as a quick shorthand there are dangers in oversimplification and care is required.

So far this book has concentrated on developing the data model to the finest degree of detail. But it is sometimes necessary to produce high-level data models for presentation to senior management or to facilitate a broad-based data analysis exercise where it is necessary to break-down the task into smaller units for detailed modelling. Such models are often known as Quick and Dirty Models. They will show the main entity types and relationships but little detail. It is quite permissible to include Many-to-Many Relationships and unresolved Recursive Relationships in such models.

A typical Quick and Dirty Model, based on Figure 6.3, is shown in Figure 6.4. Note that not all the entity types have been shown and where entity types were used to break up Many-to-Many Relationships, such as VEHICLE ON COVER, these have been replaced by a direct Many-to-Many. This data model does not tell any lies but is far less informative that the more detailed Figure 6.3. Nevertheless the technique can prove useful if the analyst wishes to provide an overview of the subject area.

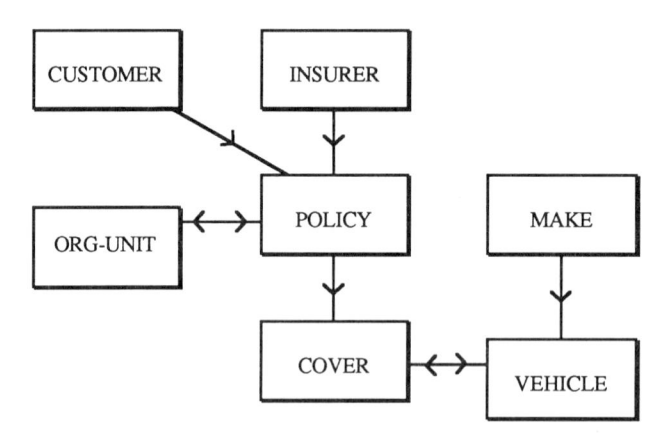

Figure 6.4   A 'Quick and Dirty' Data Model

Generally speaking, these models can be produced very quickly and with only a basic understanding of the subject area. Where the subject area is large it is useful to produce a Quick and Dirty Model and then use this to break up the area into smaller areas for detailed study, using the Quick and Dirty Model to coordinate and tie together the detailed models as they are produced. These ideas are developed further in Part 4, Methodology.

There are dangers inherent in this approach. They stem from the very fact that a model is produced at such a high level. The data analyst can never be sure that he or she has correctly understood the underlying concepts and there is a tendency to make the detailed model 'fit' the overview model thus perpetrating the misunderstanding. Such overview models should be used for guidance only and not be taken too literally.

Chapter 7

Back-Up Documentation

Purpose of the Documentation

The documentation is used to show much additional detail.

This is needed to:

- *define the entity types;*
- *identify the attributes;*
- *give examples;*
- *give volumetric and other information.*

This chapter will consider what additional information is required to aid the documentation of the data structure in addition to the data model diagram. Several situations where the conventions used in data modelling give an incomplete or inadequate view of the data in the model have been mentioned, for example where an entity occurrence can only participate in one of two or more relationships that the entity type may participate in, where membership of one relationship is exclusive of the other.

Primarily though, further detail is needed concerning the entity types in the model. It is necessary to detail the attributes that the entity type has and the identifiers of each entity type. To avoid later confusion it is also necessary to define precisely the entity type so that the reader knows exactly what is meant when a particular name is used to describe an entity type. Sample occurrences of each entity type should be given at this stage. Typical values for each of the attributes should be shown so as to provide an avenue for easier understanding of the entity type. It is also useful to gather further information on volumes and on the responsibilities for maintaining the data once it has been gathered. The back-up documentation for each entity type should be presented in the same numbering sequence as that used on the data model itself. This allows a consistent and logical presentation to be maintained.

This book does not lay down a hard and fast format for holding this back-up information. Rather, it discusses in the remainder of this chapter the information that needs to be gathered and leaves it up to individual practitioners to find their own most satisfactory means for documenting this. Remember, though, that the documentation will be revised many times as the model is developed and some thought must be given to the ease of updating. If possible, some automated method should be used either via word processing, or preferably a data dictionary if a suitable one is available.

Sample data model documentation is given in the Appendix. The reader may find it useful to refer to this whilst reading the following sections.

Content Overview

Gather the following details:

- *definition of the entity type;*
- *give examples of attribute values;*
- *volumetrics – volume, growth, retention, relationship populations;*
- *source of data;*
- *responsibility for maintenance;*
- *allow room for queries.*

The type of information which needs to be gathered can be summarised into several sub-groups. This detail needs to be obtained for each entity type on the model and is best documented by entity type.

1. *The entity type must be described and defined* to say precisely what it means. This is best done by means of free form text and the format and length of this is discussed in the following section.

2. *Examples of the entity type* should be set out showing for each example, realistic values for the attributes. This serves several purposes. It allows specification of the attributes and allows other comments to be made specifying the roles that each attribute may play within the entity type.

3. *Gather volumetric information*: not only are estimates of basic volume required, but also growth rates, retention periods for each entity type, and populations for membership within each relationship. Much of the volumetric information will be of critical importance if a computer system is implemented and it is as well to gather it as soon as possible in the analysis phase to ensure that the detail is not forgotten.

4. *Consider the source of the data* and the responsibility for updating this data. It is very important to determine where each piece of data is coming from so that appropriate arrangements can be made to gather this data prior to implementation.

   Allocating responsibility for maintaining the data at this early stage helps to discipline users into considering their responsibilities towards the end system.

5. *Allow a section for unresolved questions* or queries that still need to be answered. Allowing room in the formal documentation for

such things prevents these being mislaid or written on scraps of paper or whatever. Also as these points will be brought to the attention of each reader, the chance of obtaining accurate answers is increased.

The following sections consider each category of information in more detail.

The Description

When writing entity descriptions, bear the following points in mind:

- *take care with abstract terms, such as ACCOUNT;*
- *detail any 'upward' relationships, especially where these are optional;*
- *identify any constraints that apply to 'downward' relationships;*
- *use business terms.*

The description and definition of an entity type can be one of the most difficult parts of documenting a data model. How, for example, does one succinctly define an ACCOUNT? Everyday concepts that are not concrete like a CAR or PART, can be extremely awkward to pin down and although most people have a workable notion as to what the concept of ACCOUNT represents, this may not be satisfactory for our purposes. One must do the best one can and consider carefully the phrasing being used.

The length of the description is also worthy of some attention. It should give sufficient pertinent detail to ensure the reader understands the concept but should not be unnecessarily long and pernickety. Normally, a few sentences will suffice but for some more esoteric entity types it may be necessary to stretch to several hundred words. *In general a short, precise definition of the entity type is required.*

The description must however, give detail concerning membership in relationships as appropriate for that entity type. This is primarily concerned with 'upward' relationships, that is where the entity type being described is at the 'many' end of a relationship. It should be made clear which relationships are mandatory (One-to-Many Relationships) and which are optional (Zero-or-One-to-Many Relationships). If there are any constraints on participation in relationships (for example, exclusive relationships, which have not been made clear in the model) then these must be emphasised in the description. 'Downward' relationships require less emphasis here as the condition 'Entity Type A must have zero, one or more associated occurrences of Entity Type B', must always be true. However, if there is a particular constraint, such as 'there may not be zero' or 'there are always three', then this should be made clear.

Most importantly the entity type should be defined in terms of the business. The data model is produced at the business level and the aim is to convey this to the management of the organisation. The definition, therefore, must be in terminology that will be understandable to the users and will allow them to understand the data model as quickly and as easily as possible.

Examples

Well chosen examples aid understanding. Note these points:

- *make examples coherent between entity types;*
- *state if the examples are exhaustive or illustrative;*
- *give complete lists where the number is small;*
- *note if the examples are real or hypothetical;*
- *use real data if available;*
- *note the identifying attributes.*

It is extremely useful to provide a table of examples to illustrate the description of the entity type. This will detail and name all attributes that the entity type has and also allow typical values to be shown. The sample occurrences should be picked with care to illustrate the range of values that may occur for each attribute and to give some indication of the degree of variance or similarity between occurrences of the entity type. Furthermore, where the entity type is defined in terms of other 'higher' entity types, it is useful if consistent examples are used. For example , if 'E-type' is used as an example of MODEL then 'Jaguar' should previously have been quoted as an example of the MAKE entity type. In this way, the documentation becomes much more coherent and hangs together as the reader progresses through the descriptions.

It is often also useful to indicate whether the table given is exhaustive or illustrative, real or hypothetical, as additional guidance to the reader. Generally, if there are fewer than 20 occurrences of the entity type then it is worthwhile presenting all occurrences in the table. However, where numbers are large this is obviously impractical and an illustrative set of examples will need to be given. The only exception here is where a full specification of the occurrences would clearly aid understanding of the meaning of the entity type, in which case an exhaustive list should be given even if there are 50 or 100 or more occurrences. The table should be noted as exhaustive or illustrative.

Wherever possible real examples should be used to illustrate the table but in some cases these will not be available. In these situations examples should be invented and the table labelled as hypothetical to show that the data has been invented. If real data later becomes available then this should be substituted and the hypothetical data discarded.

Finally, the identifying attributes should be highlighted so that the reader can appreciate how each entity occurrence is uniquely identified.

Volume

Give as valid an estimate of volume as possible, but allow this to be checked by:

* *stating the derivation;*
* *stating a time factor for high-volatity entity types.*

At some point, accurate and thorough information regarding volumetrics will be needed in order to design the eventual implementation. Although this information is not required as a critical part of the data analysis exercise, it is rarely too early to start to collect such detail. It is very convenient to gather this while conducting data analysis and it is also convenient to record it along with the entity descriptions and examples. The next few sections describe the types of volumetric and statistical information required, explaining the purpose of collecting that information and its usefulness.

Firstly, volume. One of the many advantages of database systems, and a function of their flexibility, is their tolerance of fluctuating or imprecise detail concerning volume. Obviously, the more accurately a designer can size the database, the better the implementation will be. But there is *one* thing we can be sure of: if the implementation is successful, then the database will grow anyway. Certainly at this stage it is only necessary to get a feeling, an order of magnitude, for the number of occurrences of each entity type. If we estimate 10 and there are 15, or if we estimate 500 and there are 600, no one is going to lose any sleep. However, there would be concern if we estimated 10 and there were 600!

Due to the fact that the analyst will, of course, continually be referring the data model and entity descriptions to the users, this approach can be used gradually to firm up the estimate of volume, and this also allows the analyst to 'guess' at what the volume should be.

There are right and wrong ways of doing this: to state a bland estimate, of say 10,000, is not sufficient. Users will tend to assume the 'professional knows best' attitude and let it pass. Far more usefully, the estimate can be expressed functionally, based on some other factors normally derivable from the relationships in which the entity type participates. For example, if it is suspected that there are 10,000 ACCOUNTS and each must be owned by a CUSTOMER, with an estimated volume of 5,000, then this can be expressed as a useful statement: if there are 5,000 CUSTOMERS and each has an average of 2 ACCOUNTS then the volume is 10,000. This is much more meaningful. A user can pick on the statement and check its validity in a number of ways. In this case, either the number of CUSTOMERS may be queried, or the average may be deemed to be incorrect. Either way the estimate will be improved. Additionally, because the derivation of the volume has been given, should other conditions change, there is an increased chance of recognising the dependence of one estimate on another and ensuring that the appropriate revision takes place.

For most entity types this technique will be adequate. But some entity types are highly volatile and a straight number, even with its derivation, will be inadequate. In these cases, it is necessary to express the volume complete with a time factor, and state, for example, '100,000 new DEALS per year'. This latter point crosses the boundary into estimates of growth, which is covered in the next section.

Growth Rates

A clear statement of growth is required:

• quote a meaningful time factor relevant to the entity type;
• give the derivation if possible.

Although closely allied to Estimated Volume, growth rates must also be considered so that a feel for how quickly the database is going to expand can be developed. Once more, this is not central to the data analysis exercise, but an estimate of growth can be submitted to the same process of revision in order to improve and develop understanding.

In order to put growth rates into the correct perspective, it is essential that a time factor is quoted wherever possible. The choice of this time factor need not be uniform for all entity types in the model but should be picked to be as meaningful as possible for the growth rate of the entity type. Take an extreme example: if ten occurrences of an Entity Type are created on a Friday, but never any on Mondays through to Thursdays, then 'ten per week' is more effective than 'two per day'. In this extreme, where the creation is so uneven, it would also be worth noting that creation occurs exclusively on Fridays to qualify the growth statement.

Just as with volume, it is not always possible to arrive at an absolute statement of growth, especially in the earlier stages of analysis. This should be estimated where this is uncertain but, once more, this estimate should be checkable by showing the derivation. To take the CUSTOMER and ACCOUNT example from the previous section, if it is not known how many ACCOUNTS are created per year then those factors that are known should be given. For example, if each CUSTOMER has an average of 2 ACCOUNTS and the growth rate of CUSTOMERS is known, then the appropriate growth rate for ACCOUNTS can be calculated. Once more this leaves the door open for users to challenge the assumptions that have been made and provides as many opportunities as possible for them to improve the quality of the work.

Retention

Retention has to consider three factors:

* *how long is data required on-line?*
* *how long is data required in total?*
* *check for inconsistent retention periods between related entity types.*

So far this chapter has considered how many entity occurrences of each type there will be and how fast each entity type will grow. It is also necessary to consider how quickly entity occurrences will disappear or, in other words, what the retention requirements are. It is only by considering the retention requirements that a net growth rate can be determined which will eventually be a critical factor in subsequent design and physical sizing.

There are three aspects of data retention that must be considered. The first two are time-orientated; how long it is necessary to hold data in two distinct ways, the third is more fundamental to database design. The impact of retention on physical design is discussed in Part 5 of this book.

With the growth in on-line systems, the designer now needs to consider for how long data needs to be accessible instantly at the press of a button. Once more, the people in the best position to supply this information are the users – they will be the ones who are affected most critically by the availability of data or the lack of it. However, the volume of data held on-line can make vast differences to the speed of the implemented system and, where this volume is going to be large and if rapid responses are required, then this is crucial to physical design.

Secondly, it is also important to arrive at a figure for how long data needs to be retained at all. Once the immediate on-line requirement has been satisfied, will it be necessary to retain the data for a longer period in order to ensure that queries can be adequately satisfied?

Obviously, there are a number of factors here which are not relevant to data analysis – the trade-off between response time and data volumes, and the precise techniques for stripping off data from a database and yet maintaining it in a usable state are complex issues – and it is not intended to pursue them further at this point. It is still essential to develop an understanding of the requirements as early as possible. This will ensure that these requirements are considered from the outset and not cobbled together after the system has been up and running for sometime and starting to crack at the seams.

Thirdly, there is an often-ignored but very important interrelationship between the retention requirements of different entity types. To take the CUSTOMER and ACCOUNT example used earlier, if ACCOUNTS must have related CUSTOMERS, then the retention requirements on CUSTOMER must specify at least the same period as the retention requirements for ACCOUNT. If CUSTOMERS could be deleted after six months' inactivity, and ACCOUNTS one year after the last account movement, then there is clearly a logical inconsistency. It would not be possible to delete a CUSTOMER and retain his ACCOUNTS as this would be meaningless. The CUSTOMER must be retained for at least as long as the ACCOUNT details are required. By ensuring this kind of consistency and relating each entity type's retention requirements to those of the entity types to which it is directly related, it is often possible to throw up inconsistencies that can be put forward for resolution.

Finally once more it may be necessary to express the retention in relative rather than absolute terms. This is quite acceptable providing that the details of its derivation are made clear so that it can be reviewed and controlled in the light of other changes.

Average Population

Four questions must be asked about each relationship:

- *what is the maximum number of entity occurrences in each occurrence of the relationship?*
- *what is the minimum number?*
- *what is the average number?*
- *how may occurrences will be empty?*

The final area necessary to consider under volumetrics is 'average population'. So far we have concentrated on gathering the statistical information in absolute form, but the whole ethos of data analysis is that items of data are interrelated and additional details must be gathered about these relationships. The aim of this is to quantify the 'many' in each One-to-Many, or Zero-or-One-to-Many, Relationship. This detail is again critical to a successful implementation and it is as well to start considering the question at this stage.

There are four things that must be quantified. First, the maximum value for the 'many' for each relationship should be quantified. This is not always an easy question to answer. It can be difficult to put a figure on the likely maximum in situations where there is no theoretical limit. In these cases a derivative statement should be used stating what has been assumed to be the maxima, with the rider that a maximum should not be imposed. If database techniques are used anyway, it is unlikely that such a constraint would be imposed but an accurate figure for the maximum is likely to be very useful.

Secondly, the minimum value that the 'many' may take must be considered. Can it be zero? If not, what is the least number of related entity occurrences that there may be? This detail is particularly useful for qualifying the relationships that are shown. If there *must* be one or more related occurrences then it is important to say this since it is a constraint on the data in the model. Again, the practical minimum may vary from the theoretical minimum. If there are always at least, say 10 at the 'many' end except in a very small number of cases, then say so, but also say that there may be fewer.

In addition to establishing the minimum and maximum, it is useful to give some idea of the average number. This information will dominate the calculation of message path lengths and response times, disc utilisation and a number of other critical performance-related questions when, and if, a computerised system is designed.

Finally, it is also extremely useful to quantify the number of times where 'many' equals zero. In other words, how often will there be no entity occurrences at the 'many' end of the relationship. Of course, if the minimum permissible value is greater than zero, the question will not apply, but in most cases it will and the estimate can be made. Again, this detail is useful in systems design and can shed additional light on the function of the relationship.

Source of Data

Establishing the source addresses two questions:

* *is the data real? If it is, it should exist somewhere already;*
* *who understands this data?*

This section is concerned with what may again appear to be an implementation-biased question and somewhat premature for the data analysis exercise: namely, what is the source of the data initially required to build the system? To a certain extent this is true. The main use of this information will be for designing and running the system build programs and determining what data is readily available and what data needs to be gathered from user departments or wherever.

But it means more than this. Firstly, if a given entity type is included in the model, this states quite categorically that it is something of importance to the organisation under study. If the data is important, then it must already exist in some form or other and it should be possible to make a clear statement of this. Difficulties in establishing the source suggest that the 'data' may not be real but may have been contrived to meet an apparent requirement that does not exist. In this way then, establishing the source of the data ensures that only valid, real entity types are used.

Secondly, the source can be used as a pointer for obtaining further details concerning the entity type. It tells us where to look, or who to ask if there are questions that need to be answered. If the source of the data is a department or person within the organisation, then they must understand that data and its purpose.

Maintenance Responsibility

By defining maintenance responsibility, the users will be encouraged to accept that they have a clear ongoing responsibility.

Allied to establishing the source of the data, it should be possible to establish who will be responsible for maintaining the data after implementation of the system. This may not be the same as the initial source of the data and it is important to make clear what is meant by 'responsibility'. What is not meant is who is going to sit in front of a terminal creating and deleting entity occurrences as they type. Rather, the aim is to identify who takes the responsibility for ensuring that somebody does sit in front of the terminal. This can be expressed at the level appropriate for the circumstances. It may be 'Accounts Department' for example, or 'Management Committee' perhaps, if the entity type is concerned with fundamental or strategic issues.

It is useful to document responsibility at this stage. The users will be able to advise and comment, and eventually accept, that they have a clear responsibility to look after the data in the system. Again, if the data is perceived as being necessary, then somebody has to take responsibility for ensuring that the data is maintained in an up-to-date and useful state.

Questions

The major point for Questions is style. The questions should be expressed in such a way that they encourage a full and detailed response.

The final part of the back-up documentation on each entity type is concerned with any questions that require answers concerning that entity type. It is all too tempting to try to remember all the things that need clarification, or to assume that they will 'come out in the wash'. If a disciplined approach is used effectively then these queries must be noted in a consistent way. Collecting all the questions together for each entity type is also useful as it allows concentration on them in a standard way at a given point in the documentation.

The style of the questions is important. The questions themselves may range from queries on the meaning of an attribute, to fundamental questions concerning the structure of the data itself. Throughout, though, it is important to use a style which will encourage the reader to comment and respond to the questions. The question "is this entity type needed?" does not necessarily excite a user, whereas something along the lines of "do we need to hold this data on Customer in order to relate............?" is potentially much more meaningful to a user and puts the question into context. Thus, it is important to ask questions in the right way and at the right level. Questions should be asked wherever there is doubt in the mind of the data analyst even at the risk of seeming naive. It always gives users a nice warm feeling if they can show that their knowledge is superior to yours, and it will encourage comments rather than stifle them!

With a clear data model and Entity Descriptions covering all the points in this chapter, the documentation of the data structure should be clear and unambiguous. Organising the documentation in this way will facilitate its development and use as described in the next chapter.

Chapter 8
Developing a Data Model

Overview

Using an iterative method allows understanding to be progressively developed to greater levels of detail and encourages user involvement.

This chapter is concerned not so much with data analysis, but rather with a means of developing understanding and knowledge in a coordinated and consistent way. It is a convenient coincidence that this particular approach is extremely suitable for use with data analysis. The technique is reiteration, sometimes referred to as iterative design.

Why should this technique be used and how is it applied to data analysis? Throughout the earlier part of this book it has been suggested that the results of applying data analysis techniques should be continually fed back to user representatives in order to get their opinion on the quality or correctness of the model. This is in essence iterative design. This approach accepts that it is very difficult to get a full understanding of an area in one full interview or in one stage of information gathering. It is necessary to gain understanding in a series of stages, isolating the major concepts initially and gradually developing a more and more detailed appreciation of the subject area. Drawing a data model and producing a series of entity descriptions is a very powerful way of encapsulating the current understanding of an area. The data model can be discussed and debated more readily than page after page of text and the diagram can be redrawn or otherwise manipulated relatively quickly. Additionally repeated references were made in the previous chapter to the resolution of questions and firming up of estimates of volume, growth and so on. This requires an iterative approach.

One of the major problems with traditional design techniques is that users have not been actively involved in the design process. Reiteration makes this possible. Following the first series of discussions it is feasible to go back to the users with documents showing the current understanding and ask for improvements and corrections.

The technique provides a forum for each area to be gradually progressed and then documented. This helps to ensure that the area is correctly understood and that things are not misconstrued The analysis will be corrected by the users before it has gone too far off the straight and narrow. So the basis of the iterative approach is to:

gather initial information,

document it,

check it back at source,

gather more information,

document it,

check it back,

gather more information, and so on.

There are also a number of other details to be considered. In particular, how often to reissue documentation, how to control the process, the outputs of the analysis, and problems associated with ensuring consistency during the process. Each of these areas is covered in a section of this chapter.

How Often?

The frequency of issue depends, on a number of factors:

* *the breadth of the area;*
* *the number of users involved;*
* *the complexity of the area;*
* *the degree of structure within the business;*
* *the analyst's level of understanding;*
* *the method of holding and updating the documentation.*

"How Often?" This question is frequently asked, normally in a tired voice and with a pained expression! The iterative approach is often entirely new even to seasoned analysts and the thought of covering apparently the same ground over and over again does cause a sense of bewilderment.

Never fear! It is not that bad. Think of it in the same way as decorating a room. For each stage of the process – preparing the old surfaces, emulsioning, paper hanging, gloss painting, etc – the same 'ground' is covered – the walls, ceiling, etc – but each stage adds an extra dimension gradually working towards the finished product.

The iterative process is essentially the same. Having obtained a broad appreciation of the area under study, we return, review, and question, gradually adding to the detail and, more fundamentally, to our understanding of the business. The more complex the area, the harder it becomes to absorb the detail and it is necessary to increase knowledge gently, step by step.

So to return to the question of "how often?" It is essential to keep the users informed of progress. They must be encouraged to comment and review the work produced and these responses must be obtained at the appropriate time. If they are given too much detail, they will be unable to absorb what is being said; too little, and they will feel that progress is unsatisfactory and their time is being wasted.

In general terms then, it is necessary to gather information across the whole area. This should be done at a level that does not cause confusion with unnecessary details but significantly broadens understanding. Next, the documentation can be updated to include this information. If the model has been significantly improved but still maintains a resemblance to the previous version, this should be issued to the users and a review meeting arranged to discuss the model and gather their criticisms.

From the above, you will have realised that there is no fixed interval for repeating the process which can be said to be ideal. It may be anything from a week to two months depending on a number of factors. However, in the real world it will often be necessary to give some indication in advance of how frequently this repetition will occur in order to allow user management to set aside the resources required and gauge their likely commitment.

So what are these factors?

The first two are highly interrelated and will be discussed together. These are the scope of the project and the number of users involved. The relationship between these two is obvious, the larger the project the more users there are likely to be. Each additional user will bring his own viewpoint and his own requirements to the table and these will all have to be reconciled and agreed. Reconciliation of different views is important. If it takes, say half a day to interview one user, it will take two and a half days to interview five. But if it takes another half a day to absorb that information, it could well take in excess of two and a half days to reconcile five different views. The effect then, is somewhat exponential, depending on how different the various views will be. This factor is critical in determining a reasonable review period. The size of the project will have a similar effect. This increase will also be exponential.

The wider the subject area, the greater the chance of contradictory requirements and more entity types and relationships will need to be considered. Once more there is a critical trade-off here. An area that is too large will be unmanageable, an area too small will not be meaningful in itself and the results will serve no useful purpose. More guidelines on choosing project areas will be found in Part 4 of this book so the topic will be left at this point.

The next factor to consider is the complexity of the area as this will affect the speed at which change can occur. This really covers two things. Firstly, how complex, in absolute terms, is the business? Is it already highly structured with clear rules and functions, or is it extremely flexible and dynamic responding to situations and opportunities as they arise? There is a rule-of-thumb here which is useful: in general, if the business carries out a large volume of similar functions, then the area is likely to be highly structured; if the volume is low, the flexibility is likely to be much greater and the rules less well defined. This can be illustrated by looking at car manufacturers. A small family car will have a small number of options with a high-volume, mechanised production pattern. At the opposite end of the market, a luxury car may be built virtually to order with many different variations and permutations, perhaps even be hand-built, with each car different from the next.

The more structured the environment, the easier data analysis will be as the underlying structure will be more explicit and easier to define and document.

Secondly, it is necessary to consider one's own knowledge of the area. If the data analyst already has a good understanding of the business practices, the speed of comprehension will be quicker and more complete than if there is little existing expertise and therefore a much longer learning curve.

Finally, the updating process itself should be considered. Is the documentation held on a data dictionary or a wordprocessor? If so, it will be easier to amend than ordinary typescript. This is a deceptively important point. With the less automated methods the amount of paper-pushing involved can be staggering and very difficult to control. If such methods must be used, it is necessary to ensure that each iteration contains as much new information as possible in order to minimise the number of wasted trees!

Controlling Re-issues

Three control issues should be considered.
- *The review meeting*
  - *don't waste time on non-productive issues;*
  - *if a point is detailed but non-contentions cover it in depth outside the meeting;*
  - *circulate actions required;*
  - *arrange next meeting with as much notice as practicable.*
- *Ensure the documentation is issued to sufficient users*
  - *horizontally: to all affected departments;*
  - *vertically: to junior and senior managers.*
- *If several analysts are involved, take special care to control overlap and duplication of effort.*

After considering the factors governing how frequently the documentation should be reissued to the users, it is necessary to establish how to control the process of iteration. There has already been mention of some of the control aspects but they are all pulled together in this section.

The most important point to stress is that of controlling the review session at the end of each iteration. It is essential that this is effectively handled and that the maximum amount of feedback is obtained. There is considerable discussion later in this book concerning user representatives and the advantages and disadvantages of the various ways such sessions can be run. The main feature here is to ensure that discussions do not get sidetracked down blind alleys that are not going to reveal anything useful. If a point cannot be resolved, then someone should be asked to investigate the problem and provide the required information. Similarly, if a point is non-contentious but very detailed, it is more fruitful to ask someone to pull the detail together in a coordinated manner and supply it later rather than to waste valuable time covering each point of detail in discussion.

It is useful to circulate a note following meetings, mentioning the main points discussed and any action points or unresolved problems, then all are clear as to their responsibilities. Most users will be willing to come half way to meet you, but they must be sure of what it is you want from them. This is best handled by issuing a written statement on how the project is structured and the kind of detail you require. It is also necessary to ensure that they have a basic understanding of data analysis and this is best achieved by individual discussions with each user.

Allied to this it is very important to fix the date of the next review meeting as soon as possible after the last has occurred. As soon as you have been able to quantify the effort involved in preparing the next iteration, the next meeting should be scheduled allowing enough time for the next iteration to be prepared and distributed.

Having prepared the next issue, it is also necessary to ensure that a good distribution throughout the user area is achieved. This must be achieved both horizontally and vertically. Horizontally, inasmuch as it will be necessary to distribute to all departments, sections or divisions who have an interest in the study area. Sections of the company can be easily discouraged if they feel that they are being excluded from what may well be seen as an important and interesting development. This not only makes life uncomfortable but also means that their contribution will not be made and a valuable source of information could be lost. The vertical requirement is equally important. The review should gather opinion not just from senior management but also from lower levels in the organisation. Users in the lower management ranks will be most in touch with what actually happens 'at the sharp end', whilst senior management will be more concerned with strategic and policy decisions and the information needed to support these. Both viewpoints are equally valid.

Last, but by no means least, if the project is a large one involving a number of data analysts, it may well be necessary to ensure that their efforts are carefully coordinated. If the project can be split into separate areas with little overlap in terms of both data *and* users, then this will not be necessary. However, where the overlap is high, the "left hand must know what the right hand is doing". If the same users find themselves discussing the same things with a number of different data analysts they will quickly become uncooperative.

Similarly there is no point in two people independently analysing the same data structure, albeit from different viewpoints. Their efforts should be coordinated to ensure that both viewpoints are considered as a whole and the work is only done once. These may seem obvious points but it is surprising how easily they can be forgotten once the project is under way!

Outputs

The major outputs will be a revised data model and entity descriptions. The following should be noted:

- *date each version of the diagram;*
- *review the entity descriptions thoroughly;*
- *circulate minutes, action plans, etc, as soon as practicable;*
- *keep the users informed.*

This section considers what 'outputs' are generated or re-generated via the process of reiteration. These have been mentioned in passing but there are a number of points to bring out here.

Firstly, it is almost invariably the case that the data model itself will require some change, and that a re-drawn model will have to be issued. The problem of ensuring consistency between versions of the model is dealt with in the next section, but here it is useful to consider a couple of points to be borne in mind when redrawing the diagram.

The first may seem a small point, but it is extremely useful to date each model immediately it has been drawn. This is especially relevant during the later stages of data analysis when the changes become far more subtle and the shape of the model begins to stabilise. It can be quite difficult sometimes to recognise the latest version without some straightforward means of doing this. It also makes it very easy to check if somebody else has an up-to-date copy. Versions of the model can also be easily filed in chronological sequence to give a picture of how the data analysis has progressed and the areas that have been improved.

The second point concerns the numbering of entity types on the diagram as discussed in Chapter 4. It makes life a lot easier if the model is drawn in such a way as to minimise the effect of re-sequencing the numbering. If the entity descriptions have been numbered in the same sequence (and it is advisable that they should) then the less disturbance to the overall sequencing, the less re-shuffling of pages is required which makes everybody's life a lot easier. If the numbering has been done in a clear logical way to start with, this should not be too difficult to achieve, although it does take practice.

Secondly, the entity descriptions in the back-up documentation must be dealt with. If the model has been redrawn, then the back-up documentation will require an edit to support the new model. It is necessary to work carefully through the descriptions and examples, volumes and statistics, etc, ensuring that they correctly reflect the additional knowledge gained. The Questions section for each entity description will need revision, hopefully with many of the questions now resolved, but almost certainly a number of new ones to be added.

Additional outputs that would be generated are of a lessor nature and were discussed in the previous section. A report or summary of the last review meeting should be distributed to all attendees *and* other interested parties, complete with a list of action points, and the responsibility for resolving them. At the same time, or shortly afterwards, the date for the next review should be set and an agenda and details circulated to the attendees. Obviously, the amount of formality in such notes will vary depending on the size of the project and the general style of communications in the organisation: there are no universal guidelines for this.

The foregoing should be regarded as the normal minimum requirements in this area. Again, depending on individual preference or company style, it may be necessary to produce other reports or documentation. If this is what your company expects, then do it! The aim is to communicate understanding and to collect as much support and additional information as possible.

Checking Consistency Between Issues

A number of techniques can be used to prevent or detect errors:

- *count the relationships for each entity types and reconcile differences;*
- *check that each entity type on the old version is present on the new version, or justify its absence;*
- *check that attributes have not been lost;*
- *check that all relationships are supported by attributes and vice versa for supporting attributes;*
- *check that the textual entity descriptions are still valid and complete.*

When dealing with a data model of at least moderate complexity, it is extremely easy to introduce errors and make ommissions during the iterative process itself. Once the model begins to exceed about 50 entity types it seems impossible to prevent errors from creeping in – no matter how careful and methodical one may be! I'm afraid the only salvation is to check painstakingly each entity type and its relationships and justfiy each change that has been introduced.

However, in an effort to ensure that this is carried out with a modicum of efficiency and without too much tedium, there are a number of specific things to be checked that should catch the majority of problems. Some of these may appear initially obvious but it is simply because these problems are so common that it is necessary to make these points so explicit.

First of all, relationships. It is far too time-consuming to check that each realtionship has been re-drawn correctly. A quicker way of doing this is to simply count how many relationships affect each entity type. By counting these on each version of the model it is then apparent if there is a difference. If there is a mismatch in numbers then one has to fall back on checking each relationship individually to resolve the differences.

Now I can almost hear you saying, "what if I have the same number of relationships, but I have related one to the wrong entity type?" It is a valid point, but it is not too drastic. This should be highlighted when the other entity types are checked. It is possible that two or more compensating errors have been made, and if this is the case, the counting check will not detect them. Again, there is no need to worry too much, there are other ways of detecting this problem which will be dealt with later in this section.

Next, attention should be given to the entity types themselves. The concern here is to ensure that no entity types have been missed out as the data model was redrawn. (The chances of having invented some entity types without any reason are so small they are not worth worrying about.) The easiest way to achieve this check is to take the 'old' version of the data model and work methodically through it, from top to bottom and left to right, simply to ensure that each entity type on the 'old' model appears on the 'new' version. Do not worry about relationships here; as long as the entity type is there somewhere, then any problems with relationships can be resolved with the other techniques. If an entity type is found to be missing, then this has to be resolved. Clearly any missing entity types will have arisen as a result of deliberate removal, a change of name, or an oversight.

Thirdly, the Entity Descriptions need examination to check that the right attributes have carried forward. Again, as with relationships, a simple count of the number of attributes is sufficient at this stage. If there is the same number of attributes then there is no problem, if there is a difference it needs to be resolved.

Having done this, there has been a basic check that all entity types, relationships and attributes are as they should be. The next stage is to be sure that the new version of the data model is consistent in itself and to this end there are two basic cross-checks that need to be carried out.

The first of these is to ensure that each relationship shown on the model has been supported with appropriate attributes in the examples in the Entity Descriptions. There should also be a check that any supportive attributes shown are reflected as relationships drawn on the model. This will discover any misplaced relationships which have hitherto gone undetected. There is no real short cut to this. Each entity type has to be considered in turn, preferably in the numbered sequence of the new version, and the tie-up between attributes and relationships examined. Remember, though, that the error can be in either direction. The relationship may simply not be there, or it has not been shown or supported properly. It is necessary, therefore, to consider fully the implications of each mismatch.

The second cross-check is one that is especially easy to forget, yet it can make an enormous difference to the readability of the documentation. This is concerned only with the new version of the Entity Descriptions. Its aim is to make certain that the textual description and definition of the entity type still hold good against the examples and attributes shown. Where an entity type has been defined in terms of other entity types, is this still valid on the new version? Any additional detail concerning the relationships will need to be examined. Does it still apply? Is there further information that should be added? Does the description still reflect the current understanding of the entity type? Something may still have the same name and yet the understanding of that entity type may have been greatly revised. It is so very easy to forget to amend the text. After all, we know what we mean, don't we?

This chapter has considered the documentation to be produced and the process for reflecting increasing understanding of the subject area. Where the information comes from in order to produce the data model has not yet been considered. This is examined in the next two chapters.

Chapter 9

Using Existing Systems

Useful Sources

Apart from the record layouts of existing systems, there are other useful sources of data:

- *business descriptions, which are often found within system design documents;*
- *code listings and similar tables – often available as appendices;*
- *the development staff who will often have a good understanding of the business area.*

It has been repeatedly stressed that the philosophy inherent in data analysis is essentially implementation-free and is biased to concentrate on the business data required. However, most large organisations have already made a significant investment in computerised systems and it is often possible to extract much meaningful data from these systems. Remember, the desire is to concentrate on the data requirements because these will be more stable and enduring than any particular process. All systems *must* have considered their data requirements to some extent if they are of use. It should be possible, then, to extract this information from an existing system design and build it into the data model.

There are, of course, many dangers in blindly taking data from an existing system, but providing there is caution in the approach and an awareness of the main pitfalls it should be possible to make good use of this technique.

Where does one look for the data that is needed? Ideally, as much data as possible should be extracted from the existing documentation where this is of an adequate standard and up-to-date. But not all types of documentation will be greatly relevant. There is no purpose, for example, in wading through pages of detailed program specifications – at least, not yet! Possibly the two most useful forms of documentation are Record Layouts and Business Descriptions (or Functional Specifications). The usefulness of record layouts is discussed at length in the next two sections of this chapter. However, there are a number of points to be made now regarding business descriptions.

The greatest single advantage that business descriptions have is that they tend to assume only a modicum of prior knowledge by the reader. The result is that they have to explain things in common English and use less jargon. Additionally, the business requirements will be somewhat more stable than transient processing systems. The process of making a loan, for example, has remained largely unchanged for 200 years or so although the systems for administrating the process have undergone a complete transformation. Therefore, this will give a more stable and enduring description which will be a useful 'untainted' view of the business.

There are other forms of documentation that can be used. Any data processing system specifications may well be useful if they contain a reasonable amount of implementation-independent description. There is another type of information which is often hidden away in systems specifications: the sort of things often to be found in appendices at the back of the specification. Code tables and other lists of related values can be of great benefit – not just as a source of examples to feed the Entity Descriptions in due course, but as a useful prop to aid understanding. A comprehensive list of codes will yield a lot of information about the business and how it classifies and organises the world.

Lastly, do not ignore those hard-pressed colleagues in the applications development area. There will often be a large store of information concerning the current system and the business requirements tucked away inside the heads of the people who have lived and breathed with the system in the past. It is worthwhile talking at length to DP personnel where they have a thorough knowledge of the system. One must be very careful though, as many myths and misguided views have been propagated by word of mouth.

A further difficulty which may occur is that people can get so close and so involved with a system that they may not see around the problems and constraints built into that system. The analyst must be satisfied that there are sound business reasons for these constraints.

Using Existing Record Structures

There are a number of dangers to bear in mind:

- *are the current data groupings still useful?*
- *look out for repeating groups;*
- *check for redundant data;*
- *search out embedded codes, check their values and establish usage;*
- *is the data duplicated across systems and how is this defined?*

When analysing the data held within existing computer records it must be remembered that specific constraints will have affected the design of such systems. It has typically been the situation with such systems that the designer has tended to minimise the number of different record types being used and consequently the records will tend to be relatively long and possibly very complex in structure. This is not bad design but rather the result of design considerations which, though once very critical to efficient operation, are now less of a problem with modern hardware and software.

One of the major classes of problem here is caused by continually trying to work within the original design constraints once the system is operational and is being maintained. This will result in some fields having their original purpose blurred or combined with new and sometimes contradictory purposes. So, although the apparent use of a data item may appear clear cut, it is necessary to check that the field is still used in accordance with the original designer's intentions.

Data grouping

The way in which data items have been grouped together in the record is important. The original purpose for the grouping, if there was one, may now be lost in the mists of time, but it should be possible to understand the reasoning behind most groupings. Of course, it is necessary to be wary of illogical groupings as these will tend to suggest the existence of entity types that are not real. A logically grouped set of data items will tend to suggest themselves as a possible entity type but one must be careful not to be misled by the existing system.

Repeating groups

There is one typical situation frequently found in conventional record layouts which can be dealt with in a fairly mechanistic way – a virtually trained response to the situation.

This is concerned with repeating groups of data items and cross-refers to the first rule of normalisation dealt with in Chapter 5. The data analyst will often come across these groupings and they should almost without exception be broken down into an entity type of their own, obviously related to the entity type containing the data item upon which the repeating groups depend. This will give certain advantages. Primarily, it removes any restriction on the number of repetitions of the group allowable in any one instance, but there are also other benefits which arise when the time comes to implement a solution. By removing the repeating group there is no longer in-built storage inefficiency. Traditionally, repeating groups can only be stored in two ways: either by variable-length records, which are normally operationally inefficient, or by fixed-length records with a set number of repetitions. In this latter case there will normally be much wasted space as each record must be large enough to handle the greatest number of expected repetitions. Also, should the anticipated number of repeats increase, this cannot be implemented without re-sizing and re-formatting the file. By stripping out the repeating group into a separate entity type all of these problems can be avoided.

Redundancy in the record layout

As time passes, systems change and evolve. As data layouts have been designed to suit the operation of the system, then as the system changes the data requirements will also change. There must, therefore, be a check that any data items currently held in existing systems are still relevant to the organisation's requirements. Where the data item is still required, has it been adequately maintained within the system? If the maintenance is suspect, again the need to retain that data must be queried. If it is needed, how can a valid version of the data item be obtained in due course? One last point on redundancy: where a data item would appear to be no longer necessary, do check that its slot in the record has not been used for some other purpose not given in the system documentation. It should never happen, but it does, and with surprising frequency!

Embedded codes

The dangers here are quite straightforward. First of all there is a need to establish the validity of the code itself. Is it needed, and is it correct in structure? Codes will often have been extended as the system has evolved and it is necessary to be sure that the extention of the code is in line with the original intention. Also the code, or a subset of the code values, may have been added or been used for another purpose. The problem here is that one coding structure is being used for two or more purposes. Furthermore, these values may overlap and interact which will make the unravelling of the codes considerably more difficult.

There is a further series of questions about codes. For instance, do the different codes interract not just within each code table but also between them? This interaction may not be explicit and can take two particular forms. Firstly, certain values in one table may restrict the values that can be held in another table; this form of valid combination can be easily modelled. Secondly, the combination of the values may have meaning over and above the meaning of the two codes – 'The whole is greater than the sum of the parts' – which may not be made explicit in the documentation, but nonetheless may be extremely important. Finally on code tables, one must be sure that the list of values to hand is up-to-date and that other values have not been 'sneaked' in to solve operational problems that have arisen from time to time.

Data duplication

Another problem with conventional systems is the inability of data to be shared easily between systems and processes, with the result that data often has to be duplicated between different systems and across files. Two problems immediately arise here. Firstly, if two fields have the same name in different files, is it certain that they are the same and that the values carry the same meaning and are consistent in their usage? Secondly, the converse of this must be established, where the same values or data items are used in different records or files but are referred to by different names. These situations can be difficult to detect but efforts must be made to uncover them as they can have far-reaching implications on the data model and the relationships between the entity types.

Having said all of this, let us regain a sense of perspective. This section has concentrated purely on the negative aspects of data held in existing records, whereas there will be considerable benefits to be gained. Most data already held in the existing system or systems will be pertinent and relevant to the business. Most of it will be updated and maintained adequately and any groupings of data items will have a clear logic underpinning them. All these factors can be utilised to help to establish the correct data structure with the right attributes in the right entity types and with valid and meaningful relationships linking them together. Finally, the normalisation techniques described in Chapter 5 are extremely useful when applied to existing record formats, and can quickly give structure to the extracted data items.

Implementation Dependent Fields

Different types of implementation fields require different treatment:

- *Sort keys: if 'data-free' these may be ignored;*
- *Location modifiers: these carry a strong indication of a hidden relationship;*
- *Process triggers: if these relate to a real business need, they must be modelled;*
- *Housekeeping fields: these can normally be ignored but be careful with Record Types;*
- *Derived data: if the elements needed to derive the value are covered elsewhere then it need not be modelled.*

This section consider those types of data item that frequently occur in conventional systems – and also in database systems – and yet are things to avoid as far as data analysis is concerned. These are all implementation fields of one sort or another and each shall be considered in turn.

Sort keys

The immediate reaction to sort keys is to ignore this type of data item as something exclusively invented to satisfy some process or reporting requirement. Do be a little wary of this automatic reaction. Is it clear that the sort key itself has no data value which is not represented elsewhere? If it is 'data-free', what is the purpose of the field? Is it purely used for sorting and listing? If there is no other need for the data item it can be removed.

Location modifiers and other pointers

Location modifiers and other pointers normally have a strong correlation with the physical storage of data. The data item is then pure implementation and values of the item, say nothing about the data structure. Once again this may not be so straightforward. The fact that a pointer or modifier referring to one data record is embedded in another data record implies a strong relationship between the data content of the two records. This relationship must be captured in the model. This is the most obvious clue to the existence of a relationship that we are likely to get and, as such, the relationships will probably be highly important to the business.

Process triggers

There is a further type of data item that is perhaps less obvious. I have termed these 'process trigger' fields. They are typically included in order to initiate or control some process that acts on the data records. There are two types here – implementation and (obviously) non-implementation.

The difference is critical and is determined by the type of process that is initiated. If the process is purely concerned with easing some problem inherent in the system design it can be disregarded in the analysis. However if the field is used to control or trigger a process that is essential to the business – that is, it would be needed regardless of the computer system or manual system used to service the business – then it is valid for data analysis and will need to be represented. The data model will, at the end of the day, have to support the process being initiated: therefore the data to control it must be modelled.

Housekeeping fields

Housekeeping includes such fields as record lengths, possibly record types, file generation numbers, control totals, security counts and other integrity checks. The common theme behind all these fields is that they carry no data value of their own. They relate exclusively to the current implementation and can safely be disregarded. There is a possible exception – Record Types. Record Types or Data Types may well imply different *qualities* of data held in the same record. These variations will often relate to different entity types in the structure and these will need to be modelled correctly.

Derived data

Derived data is frequently found in conventional systems to speed up processing or for some other system-invoked reason. Derived data is any item that can be calculated or 'derived' from one or more data items in the same or different entity types, or, in a conventional system, other record types. We need take no account of operational efficiency in the data analysis. If it is possible to derive the value from the elementary or raw data items, then this is all that is needed. Wherever the derived value is needed it can be calculated. This may well appear very inefficient – but this is not a concern at this stage!

Using Program Code

Program level detail gives data on the following:

- *business rules expressed as conditional statements;*
- *code tables, their usage and valid values;*
- *data vetting and similar constraints on data.*

Even in the best-run installations with the highest documentation standards available, there will still be instances where the behaviour of an implemented system (that is, what it actually does) will differ from that stated in the system documentation. In some installations there may be little or no adequate system documentation and it is only at the program level that a clear understanding of the system can actually be obtained.

Examining program documentation and program code is a time-consuming business and should not be undertaken simply for its own sake. Here a judgement has to be made. What will be the value of examining the system at this level – will we learn much that is not documented elsewhere – how good is the documentation and can it be relied upon?

If it is necessary to examine the program code and specifications there are three levels of approach. To be absolutely sure that the most detailed up-to-date view of the system has been obtained it may be necessary to actually go to the program code and interpret it. This is *not* recommended except in extreme cases! In most situations there are two other sources which are somewhat easier to understand. These are the program specification and the record of alterations to that specification held in a modifications register or equivalent change-control log. These two in combination should equal the implemented code and will be easier to understand.

This process tends to carry the analysis closer and closer to the current implemented solution and further and further away from the business problem. The current solution may be a good match for the problem, or it may be terrible. The documentation will not give any guidance here. It is necessary to continually refer back to the business level to check the validity and usefulness of the knowlege gained at this level.

What types of information may be usefully gathered from program-level documentation? There are three categories here. Firstly, it is a useful source of rules. Program specifications and the source code normally contain many conditional statements which are of interest. The concern is not with what the program does, rather, what the program checks before it does it; that is, the conditions that have to be satisfied. These rules should reflect business requirements, and should ensure that the process will only be invoked on clean data. The data analysis techniques concerned with rules were discussed in Chapters 3-5 and it is here that the raw material upon which the techniques can be exercised may be usefully obtained.

Secondly, it is possible to gather much information about code tables and their usage. It should be possible to extract a complete list of the valid settings and, if the documentation is good, the precise meaning of each setting. This is essential detail. It will help to determine if the code table is adequate and if it is a simple table or if it has been extended and compromised to handle a more complex situation. Again, the relevant techniques have already been discussed which will enable the true effect of the code table to be unravelled and the required data structure to be decided.

Finally, it is possible to find a great deal of information concerning data vetting – that is, the conditions that each piece of data must satisfy before it is accepted into the system. In many cases this will help to establish the true meaning of an item of data. If the rules that the data item must satisfy are understood this will aid a clear understanding of what the data represents even where the field is badly named or its usage is unclear. Data vetting requirements will also say much about how different data items are interrelated and what combinations of data are allowable. Again, this will allow the correct data structure to be determined, establishing the need for cross-reference entity types and the like.

Exceptions Processing and Short-cut Solutions

Look for exceptional conditions and other unusual situations. In particular:

- *the 'Not If' situation;*
- *the 'If and Only If' situation;*
- *the undocumented one-off program.*

Perhaps this section should begin with a brief discussion as to what is meant by the title. This section refers to those little quirks that can be found in any system and that, although they may be taken for granted or tolerated, can give valuable information about the real requirements of the organisation.

What is the value in examining exception conditions? Recall the short discussion on the value of negative information mentioned in Chapter 3. There the point was made that negative information, although often harder to understand, actually provides more clues to requirements than positive information. In other words, 'what is not' is far more useful to know than 'what is' even though it may cause us to reconstruct our thinking and alter our view point. Take an everyday example: the question, 'can birds fly?', would elicit a positive response, but the question, 'can *all* birds fly?', requires a negative reply. This tells us we have to apply the general rule with care and need to cater for exceptions.

The same can be said of exception processing. Those occasions when action has to be taken purely to satisfy an unusual set of conditions will often say a great deal more about the requirements than those instances that simply confirm the current understanding. The eventual data model must support any exception processing that is needed and it will require thorough analysis to expose these situations. It is precisely because they are not the norm that we may not be told about them by the users. It is for exactly this same reason that it is necessary to find out about them. Just because they are unusual does not mean that they are unimportant. Often, in fact, the converse is true.

These unusual occurrences are of fundamental importance to understanding the business and may shed a great deal of light on the normal processes by placing them in a more realistic perspective. The data analyst, then, must adopt a suitable frame of mind and an approach that deliberately questions the current understanding. One must set out to disprove current theories and actively search out those situations that are awkward and contentious. This is very difficult: we all like to think we have understood things and it is human nature to seek confirmation of our ideas rather than seeking to expose the gaps in our knowledge. Nevertheless, this is the attitude that must be taken!

The 'Not If' situation

So what sorts of things are of interest? Firstly what might be called the 'Not If' condition. This will typically be some variation of, "if A is less than 10 then do B, but NOT IF C is equal to 2." In other words, a condition which is fulfilled under almost every circumstance but very occasionally unfulfilled, when some other course of action needs to be taken. Again, it is because the condition is rarely unfulfilled that the consequences may not be highlighted. Users may not even realise that this information is important, and it may not be documented anywhere. So beware: if B apparently always follows A, the question is, "does it..........?"

The 'If and Only If' situation

The next point to consider is the 'If and Only If' condition. This is a similar form of expression to the Not If condition although it is slightly more positive in form, implying that specific action is taken under a particular circumstance: for example, "Always do B and IF AND ONLY IF A = 10, then do C". Typically, the situation will not be made clear to the data analyst for the reasons given above. It is not that people deliberately try to deceive, it is simply that they may not appreciate the requirements!

The undocumented one-off program

It is now appropriate to shift attention to another type of exception that is often ignored when carrying out analysis; the undocumented one-off. Everyone in data processing will be aware of the 'simple little program' that gets cobbled together to solve an unusual user query and gradually evolves into one of the most important programs in the system! It does happen and what is worse, these situations may be totally undocumented and the analyst may not be made aware of them simply because they have always been regarded as a somewhat unofficial piece of programming. These need to be thoroughly investigated. The fact that it has been 'slipped in' around the existing design will also suggest that it may be a compromise of the true requirement or that it may compromise other requirements. Short-term measures, short-cuts and other 'bodges' (all of which may become long-term solutions) may often be sufficient to allow the user to tolerate the problem but may not solve the underlying difficulty. Once again, thorough analysis is required. The attitude that 'This happens rarely, so I'll ignore it', encourages complacency and simplistic solutions. You should be saying "This processing is different .... WHY?", in order to ensure that the full requirements of the system are modelled and that *all* the essential features of the business are supported.

Problems with these Sources

Concentrating too much on low-level implementation detail creates dangers in the following areas:

- *design compromises will fudge 'reality';*
- *similar but different concepts may be combined;*
- *post-implementation changes may be difficult to unravel;*
- *unnormalised data structures and derived data;*
- *risk of propagating system limitations.*

All of these add up to 'bending the world to fit the system'.

The final section in this chapter highlights the problems inherent in using this low-level of detail. The data analyst must be aware of these dangers otherwise the result will be no better than the current implementation – and this is not the aim!

Consider design compromises. As the development life-cycle moves from the original Requirements Specification through design to implementation, some degree of compromise will be introduced *regardless* of the techniques and software used. There must be an awareness of this when existing systems are used as a source for the data model. The implemented solution will not be a complete answer and it should never be assumed to be so. Findings must be checked with the users to discover where the compromises have been made.

There is also a tendency to treat unlike as like. For implementation efficiency two different concepts may be treated as similar if there is a moderate correspondence between the two. This will be apparent where exception conditions abound around the use of the same code table or categorisation. These variables must be modelled separately taking account, if necessary, of the relationship between the two.

As systems grow older they will evolve and changes will be introduced post-implementation. It is often the case that these changes are poorly documented and, because they will have been made as a direct result of an operational requirement, they may not have been fully thought through.

These short-cut solutions will need to be fully investigated. They are almost certainly essential requirements or they would not have been added. Unfortunately the lack of documentation and the urgency with which the changes may have been made, can cause difficulty in uncovering the real requirements.

This type of problem can be summed up as "What is, may not be what should be", and can be applied to any part of the system whether post-implementation or part of the original design. Again it is a matter of adopting the appropriate frame of mind and asking about the short comings of the existing system. What does it not do adequately? Where can the system be improved? These are the kind of questions that need to be asked. Development backlogs and other difficulties may have caused these problems but when carrying out the data analysis we should attempt to free ourselves of these restrictions.

Additionally, data structures used in any existing system will probably be unnormalised. As such the normalisation rules will have to be applied, either formally or informally, to break down the data content of existing files and construct a valid data model. Existing files will tend to contain much derived data (which must be analysed to ensure that there values can be calculated from existing data items) and also implementation fields used for file management or housekeeping purposes.

There is also the danger of carrying over any limitations of the existing system into the data model. Typically these will be such things as repeating groups, numbering sequences, field sizes and volumes. The repeating groups should be broken out into separate entity types, as for First Normal Form, and this will additionally remove any restriction on the number of times the group may repeat.

For numbering sequences, each should be checked to ensure that it is logical and has meaning outside the current system. Similarly, is it restrictive? Are there potential future problems even if it meets current requirements? Field sizes, although not central to data modelling, should also be considered, especially where these relate to codes or similar data. Is a two-character code adequate for future expansion? Is it sensible and clear in its range of values? Is there a better means of holding the code? The data model will not specify field sizes, but if there is a problem in this area, it must be noted in order to avoid the problem during implementation. Volumetric information may also be well out-of-date after the system has been implemented for a long period, or it may simply have been wrong to start with. Such details should be checked before they are used in the documentation. What effect will future expansion have on those volumes? How quickly is the business expected to expand and in what way?

There will almost certainly be parts of the business that are completely ignored by the current system and this should not be a constraint. The scope of the data analysis should not be prescribed by the scope of the existing system. The business requirements set the scope and any interfaces between computerised and non-computerised areas, or between separate computer systems, have to be adequately considered and the impact of one on the other fully assessed.

Some areas of course, may be badly handled even by a current implementation, so the reliance that is placed on information from a current system must depend on how well the current system meets the business requirements. This must be checked with the users: they are the only people who will give an honest answer!

A final general warning which effectively summarises the foregoing: all these problems add up to 'bending the world', or in some other way departing from the reality of the situation. However, the compromise may come from one of two sources. If the compromise has been introduced to make the system easier to implement – the situations discussed above – then the real requirements should be unravelled for the model. Compromises may also have been introduced by the business itself simply because that is the way it wishes to perceive a given situation. This may be perfectly valid and if that is really what the business wants to do then it becomes 'reality' as far as this exercise is concerned and the data model should reflect this. These compromises should be questioned, though, to give the opportunity for them to be removed if this is felt to be better in the long-term.

Much detail can be obtained from existing systems but it is also important to establish good contact with the users. This is the subject of the next chapter.

Chapter 10

Involving Users

Get Commitment

It is necessary to involve different levels of user:

* *senior management for policy and strategy;*
* *junior management for day-to-day operations.*

Additionally, senior management must be convinced of the value of the exercise in order to create the right atmosphere for the time and effort involved.

Although the last chapter discussed ways of extracting information from existing systems, probably the most important source of information for the data analysis approach is the user. It is absolutely central to the method outlined in this book to involve users as much as possible at every stage where they can usefully contribute. Whilst it would not be expected to gain user feedback on most technical design criteria, it is essential that they are involved whenever information about the needs and requirements of the business is being gathered. The users are the only source that can accurately describe two things: firstly, what the business does including its operational requirements and, secondly, where it is likely to go in the future, its policy and strategy for development.

What level of user is required? The two types of questions mentioned above require different people to provide the answers. Senior management will be aware of the policy and strategy and will be concerned with the overall direction in which the company is moving. Middle and junior management will be able, by and large, to provide detail on the immediate needs of the organisation and the processes it carries out. Therefore, it is necessary to talk to both categories of people.

There is another factor which is critical in ensuring success. This is commitment by the users to the exercise. A data analysis exercise is almost certain to fail if commitment cannot be obtained from the users.

The management will almost inevitably be busy people, somewhat unwilling to give up a lot of their time to "yet another computer thing" as they will see it. The onus is on the analyst to ensure that every effort is made to keep the users involved and interested in the project, but this does not help to get the commitment in the first place. This emphasis can only come from senior management. Management at the highest possible level must be convinced of the value of the exercise and its benefits to the company. Senior management will then be able to create the right atmosphere for discussions with middle and junior management. After that it's up to us!

The means of establishing this contact and the different ways of involving users are discussed in the remainder of this chapter.

Establish Contacts

Different types of contacts need to be established:

- *senior management will need to be convinced of the value of the exercise and their commitment obtained;*
- *middle management will be able to provide detailed operational information;*
- *end users can be a useful source of information on 'how it really is'.*

The aim is to fulfil a number of goals:

- *commitment at all levels;*
- *education in the techniques;*
- *user responsibility for the end result.*

In order to get the degree of backing required and also to obtain the basic detail needed, it is necessary to establish the right contacts with the right people in the right way. Who are these people and how should this be done?

Start at the top and work down. It is not enough for senior management simply to pay lip service to the need for on-going user support: they must be convinced of it and it is up to the data analyst to ensure that this commitment is forthcoming. So the data analyst must prepare the case and present it in the right way. There is no simple answer to how to do this. Whether single interviews or formal presentations are used will depend on the number of people involved, the availability of these decision-makers and the formality or informality of the organisation. Regardless of which approach is used the aim is the same. Senior management will want to know what is to be done, why it is necessary, the benefits of the approach over other techniques and what they are going to get out of it. Most of the arguments required have been presented in the early chapters of this book and will not be repeated here. Senior management will also want to know how much involvement they and their staff need to provide and the type of information they should make available. Senior management will *not* want a detailed explanation of data analysis, rather an appreciation. If the confidence of senior management can be won and their support and involvement at the strategy and policy level obtained then the exercise is well on the way to success. Consideration of the other user staff to be involved can then begin.

If senior management can provide the policy and strategy, then direct user management will be able to provide the knowledge of the detailed requirements at the operational level. In order to establish contacts the same techniques can be used but there is a reliance on the senior managers to specify the appropriate contacts and to ensure that their time will be made available. Be careful, apply any pressure gently. Resentment built up now will rebound at a later and undoubtedly critical stage.

If the organisation is large it may also be necessary to establish contacts with selected end-users of the eventual system. These will be the people 'on the ground' who will be directly aware of any existing operational problems and this level of detail will be very useful as the modelling proceeds.

By now you will be wondering how on earth you are going to find time to talk to all of these people all of the time. The short answer is that you will not! By establishing contact at this stage you are making everyone aware of what is happening and that you *may* need their help. It is essential to prevent parts of the organisation feeling 'left out in the cold'. Furthermore, by justifying your presence now you can save much time later when you need to interview a particular user.

At the end of this introductory stage a number of things should have been achieved. Firstly, educational: all interested parties should be aware of what is happening, why it is happening and their role in it. Secondly, commitment at various levels, to ensure that appropriate help will be available. Thirdly, responsibility: if the users do not help, the system will not work. In a very real sense it is the users' responsibility to ensure that the design satisfies their requirements. Lastly, attitude: an atmosphere of co-operation and of a common goal must be built. Users must be aware of the benefits that they will get back from their efforts at this stage. Once this has been achieved we can consider how we obtain the information we want in the way we want it and at the time we want it!

Talking to Users

It is very important to involve the users. To this end they must understand the approach. Existing data models or overview models can be used to get started and to build confidence. Do not use detailed models until the users are confident enough to understand them. 'Real-time data modelling' can be a valuable technique, but is high risk. Organise it with care. Don't patronise the users.

There are basically two ways of approaching users to obtain the detail necessary for the data modelling exercise. This may be either a One-for-One interview approach as discussed in the next section, or it may be achieved by gathering a number of user representatives together and holding a group discussion as discussed later in this chapter. However, before dealing with the relative advantages and disadvantages of each, a number of things are common to both approaches and these are discussed below.

Remember it is essential to *involve* the users. The analyst should not simply use them as a source of information and then forget them. In order to achieve this involvement it is important to establish a level of understanding between all parties. The analyst must come to understand the business and what it does, and the users must be able to understand the techniques used in data analysis. This must be done carefully and with due consideration. Do not attempt to turn all the users into data analysts but try to ensure that they do understand the fundamental concepts given in Chapter 2. What is more they must understand *why* the approach is taking what at first may appear to be a long-winded approach that requires so much of their valuable time.

It is therefore necessary to ensure that each concept is explained carefully and that the implications, especially of relationships, are fully understood. How easily all this is achieved will depend on the receptiveness of the users and the analyst's own abilities so it is not possible to give clear guidelines here. Suffice it to say that this is important and it is worth setting aside a generous amount of time to allow for any difficulties.

It is useful to employ any existing data models to guide discussion and in some cases to ensure that you have understood things correctly. Depending on the amount of knowledge you already have and the familiarity of the users with data modelling, the types of models employed will vary, and may also be usefully combined. However, there are three main types that may be used.

There may already be a high-level overview model built either from existing knowledge or from discussions with other people. If there is one available it can be used both to show how data analysis works, and also to guide and target discussion on particular areas. It is possible to talk through each entity type and relationship in some logical sequence, explaining what is currently understood and questioning assumptions in order to elicit further detail. Wherever the model implies any restriction on what may occur it is important to spell this out so that everyone is quite clear about the situation. It is dangerous to assume that others will automatically appreciate the constraints inherent in any relationship. Often, if two entity types are linked by a relationship, users will assume that the relationship is of the correct nature. This must not be allowed to go unchecked!

If the data analysis has been progressing well and you are, perhaps, in the later stages of modelling and have users that you understand, it is possible to use detailed data models as a basis for discussion. One must be careful here. To present a detailed model to a user who is unfamiliar with the techniques will have a negative effect. It can be somewhat off-putting to be faced with 50 entity types and as many relationships if you are unsure as to what all the boxes and arrows signify!

Having said that, never assume that a situation has been modelled correctly. This certainty can only be achieved by referring back to the users for confirmation. A detailed data model can be very useful here if you are sure that it will not cause total confusion.

At the other extreme there will be occasions when you are starting with a blank sheet of paper, as it were. You will probably (indeed hopefully) have an idea of the subject matter, but may not have as yet produced any form of data model. Rather than just interviewing or discussing the subject in the normal way, there is a useful technique that can be applied here. This is to use sketch models or, as I prefer to call it, 'real-time data modelling'.

The aim is to develop the model by sketching it out as it is explained. This is high risk! Before attempting this the data analyst must be confident of sufficient proficiency to cope with whatever horrors turn up. If it is employed successfully it has two distinct advantages. Firstly, it allows a clear picture of the discussion to be built up as the subject unfolds and thus provides a central focus for developing each idea. If something is drawn incorrectly it can be instantly recognised and corrected, so much time can be saved. The models so drawn also form a convenient 'hard-copy' which can be taken away as a record of what was discussed and used in the production of the documentation. The second main advantage is that it demonstrates the usefulness of the technique and actually shows 'how it is done'. Discussions can often liven up as the users see the model taking shape as they talk. They know you understand them, and you know they understand you.

How is this technique controlled? First make sure there is sufficient material. This may sound trite but it is necessary. Once discussions have begun all that needs to be done is to pick out the concepts as they are introduced and draw an entity type for each. An earlier chapter suggested the technique of getting a business description and underlining the nouns to identify the entity types. This is similar. The concepts will roughly correspond to the major nouns as the subject is described and the verbs used will *tend* to indicate relationships. That sounds very easy but it quickly gets out-of-hand. You will misunderstand things, and you will need to correct the drawing as it develops. It will therefore get very untidy very quickly. At a suitable point then, one must get a clean sheet and redraw with reasonable neatness (but don't lose the impetus of the discussion!) the entity types and relationships that have been agreed and then carry on with the development. Use simple structures whenever possible even if you know you will have to make it more elegant later. It is important to keep the discussion flowing. Do not produce works of fine art. If the sketch is clear, carry on!

All the other concerns regarding use of data models apply equally well here. Each relationship drawn will imply certain constraints on the entity types connected – make sure these are understood. But remember, real-time data modelling is risky. If you fail to understand what is explained quickly or understand it but cannot think how to model it, it can be a turn-off for the users. Your success will also depend on how well you know your users. If you have established a good relationship with them they will allow you a certain amount of muddling as they will trust you to tidy it up later.

If they are already suspicious, then don't try it!

Having stressed the necessity of explaining data analysis to the users, I must add a few words of warning. Do not overdo it. Be wary of patronising the very people whose co-operation you need most. The aim is to carry the users along in a spirit of mutual respect and cooperation. They supply the business knowledge and you supply the technique. Together you should be able to produce an accurate data model.

Single User Interviews

One-for-one interviews need careful planning:
- *develop an interview plan;*
- *pre-determine the duration;*
- *circulate discussion material in advance;*
- *consider the location;*
- *structure the approach;*
- *use probing questions.*

This is the first of the two main approaches to gathering information. The method is to carry out one-for-one interviews with the relevant users and as such differs little from traditional interviewing. In the initial stages of analysis these interviews will be used to gather information about the business and its data, but as the analysis progresses, the interview will be used more extensively to check or confirm the accuracy of the developing data model.

In both situations, adequate preparation is essential. Even when starting from scratch it should still be possible to establish an interview plan to guide the discussions. This need not be a formal agenda, but the user representative will need to know, preferably in advance, the scope and subject of the interview. An important adjunct to this is to arrange the interview for a suitable duration. This will help to keep the discussion moving so that all items are discussed. It is always possible to return to an interesting point of detail at a later date, and this is preferable to getting bogged-down in some minutiae of detail and not discussing other subjects. Fixing a suitable duration will also help ensure that the chosen representative will be able to set the whole time aside, thus reducing the likelihood of unscheduled interruptions. If necessary, it may be preferable to arrange a meeting-room or other neutral territory in which to hold the meeting, which can also reduce the chances of other business intruding into the discussion.

If the purpose of the meeting is to discuss a data model derived from earlier meetings, then do ensure that the user has been given a copy of the model sufficiently in advance for this to have been read and absorbed. Also, it is useful to make it quite clear that prior reading and digestion of its contents is a pre-requisite of the discussion.

It is also worth considering the correct interview approach. It is necessary to be more structured about fact-finding than the traditional amoeba-like absorption of information that typifies classical analysis. Try to avoid discussions of an implemented solution. The aim is to discover what data is important to that user and how it is used. A 'helpful' user will typically try to give you a solution to the problem ("We need an on-line system with a graphics capability......") rather than detail the essential building blocks that are required. Similarly, there is little value in becoming engrossed in discussions on the short-comings of any existing system. These topics must be steered back to the underlying business requirements as carefully as possible. It is also essential to ask the right questions and prompt for information. Question if data relationships can change over time, search for exceptions, and try to establish what 'rules' may restrict what can be done and how. If the discussion is left fairly unstructured the user will tell you what he thinks you want to know, and this must be avoided.

What are the advantages of the one-for-one approach? Primarily, they lie in the ability to take an in-depth look at various aspects of the business that may be of no interest to other users.

It is a relatively easy way of probing that user's problems and gaining an understanding of the business from that user's viewpoint, but there are a number of distinct disadvantages. Paradoxically these also stem from the fact that it is one person's view. That particular user's view is likely to be slightly or even considerably different from other users who are involved in the same business area and the data analyst has no immediate means of objectively determining which view is correct. It is possible to end up with several different and irreconcilable descriptions of the data and processes required to service the same business activity, and this can make life very difficult.

However, the one-for-one interview can be used where it is necessary to delve into the detail of a specific and clearly defined area. The information gained must then be incorporated with other detail and built into the data model. Most importantly, as the data model progresses, it is essential to check back with the users to ensure that what has been constructed is correct *and* meaningful.

Group Discussions

Group discussions can be difficult to organise and require considerable effort to execute well, but the extra benefits are worth the effort:

- *all views gathered in one session;*
- *time-saving;*
- *builds group responsibility.*

There are disadvantages:

- *difficult to chair, direct and understand simultaneously;*
- *control of the necessary warm-up period;*
- *hard work for the users.*

Group discussion is a rather different approach to the problem of information gathering. It is harder to organise, it requires good preparation, it is very hard work and it can be time-consuming. Nevertheless it is very useful!

This approach can be referred to as the user jamboree and is concerned with getting a number of user representatives together to discuss a specific part of the business. Thorough preparation is a must with this approach. The analyst and all the other attendees must have a clear idea of the subjects to be discussed. The meeting must be well organised with a suitable room available, adequate provision for refreshments and a generally professional approach. Meetings will need to be arranged well in advance to ensure that people will be available and pressure applied, where possible, to prevent people pulling-out at the last minute.

Given that this approach requires much extra work, what are the benefits? Above all, it is the opportunity to get a single subject discussed by all interested users in one session. Each will be able to draw on the others' viewpoints and the chances of arriving at a consensus and a correct conclusion are considerably increased. It is quite surprising how frequently different viewpoints and understandings can be resolved with this kind of open forum. It also engenders a deeper feeling of responsibility to ensure the production of a correct model. Users will be able to identify themselves as a group whose responsibility it is to provide the input for the data model; it is easier to convince users that it is their responsibility to ensure that the model is correct and that it meets all the requirements.

Although discussions take longer than they would with a single user, the ability to gather several different viewpoints together and resolve the conflicts as they arise can be inherently time-saving in the long run. Wherever there is a need to gather the views of a number of users on the same subject, and especially where their requirements will differ, the user jamboree can be used to good effect.

There are disadvantages. If the data analyst is taking the role of chairman in such discussions, it is necessary to control the direction of the meeting. It will be necessary to limit discussion where this is leading down an obvious blind-alley or moving off at a tangent. There will be situations where differences of opinion cannot be resolved within the group and it will be necessary to direct these elsewhere, normally to a higher level of authority. Also, to establish a useful discussion requires time. Each individual meeting will take time to 'warm up' and it will also take a period of time for the attendees to develop a good working relationship within the group. This can make things very difficult for the data analyst, who not only has to understand and contribute to the discussion, but has to direct and control the discussion. Further more, the speed with which the group establishes its identity will also be due, in some part, to the interpersonnel skills of the analyst. Because of the warm-up period required (typically 45-60 minutes), it is often useful to arrange such discussions on a half-day or preferably full-day basis. This can be difficult but it is well worthwhile – it also makes the meeting very taxing on all involved.

To maintain concentration over a protracted meeting is hard for most people and the person chairing the meeting must learn to detect when people are 'switching off' and take appropriate corrective action, either by calling a halt to the meeting, adjourning for a 'comfort-break', or simply cutting discussion and changing subjects.

Despite all the additional effort required, the user jamboree can be an extremely effective method of gathering information and can far outweigh the more traditional approach of one-for-one interviews. In practice both need to be used. The critical factor is how many user areas are interested in a particular subject. If the topic is only of interest to one or two user areas, one-for-one interviews will be easier to organise and will not bore everyone else to tears. However, where the subject is common to many areas, a group discussion is likely to be more effective in eliciting a valid solution.

Establish Data Usage

If a piece of data is relevant then further questions should be asked:

- *who is responsible for the data?*
- *who creates it?*
- *who enquires on it?*
- *who updates it?*
- *who deletes it?*
- *what are the security and access control implications?*

These questions are relevant not only to entity types but also to the relationships between the entity types.

Having determined how the user discussions are going to be organised, it is possible to move ahead and start to gather the information required. Obviously the primary aim is to establish the entity types, attributes and relationships so that the data model can be constructed. However, as well as establishing the base data that is needed it is also essential to consider the usage of the data. There is no point in recording a piece of data if no one wants to *use* it in some way.

What is *not* of concern in this section are the processes that affect the data. This is a large topic in its own right and is considered in Part 3 of this book. What is important is to whom the data is relevant. Where in the organisation does responsibility for this data lie?

Two distinguishable questions should be asked. Firstly, "who is responsible for the piece of data in an absolute sense?" This is to establish the area of the business that would be obliged to ensure that all occurrences of a given entity type are correct. This is not a question of who actually sits at a terminal and presses the keys but who requires that this be done. For such things as Customers and other 'external' data it may be the sales department or whatever, but, for the organisation of the company and the rules governing who does what, it may be the Directorate or some other high-level body.

The second question is equally important and is concerned with the relevance of the data to various parts of the organisation. Here the question must be broken down further. Basically there are four types of activity that may be carried out on any piece of data and the user departments responsible for each of the four may be different. It is important to establish which departments will need to *create* instances of the entity type, which departments will wish to *update* data already created, who needs to *enquire* on that data and lastly, but equally important, who has the job of *deleting* data no longer required. This information is crucial to the later stages of systems design and it is relatively straightforward to obtain it while the users are available to the analyst.

Consideration of these questions should not be restricted to entity types. Equally important is the responsibility and relevance of the relationships between entity occurrences. The relationship may be allowed to exist between two entity types, but who determines that particular occurrences of the entity types need to be related? Who uses the relationship for enquiries? Can it change and, if so, who changes it? By accepting that the concept of relationships between data is central to data analysis, then the creation and use of relationships in any given application must be equally important.

The final area to consider here is the question of access control and security. Some applications, in fact an ever-increasing number, need to restrict access to certain information for reasons of security or confidentiality. Where this is required, positive steps will need to be taken in the physical systems design to reflect these requirements.

Again, while the user representatives are available there is an ideal opportunity to probe these requirements and establish the ground rules for later work. The earlier these aspects are considered, the easier it will be to design an efficient and elegant solution rather than being taken by surprise by these requirements much later in the design phase.

Resolving Conflicts

The recognition and resolution of conflicts is an essential activity:

- *use 'round the table' discussions to elicit agreement;*
- *involve more senior managers to mediate;*
- *chase through the resolution;*
- *inform people of decisions made;*
- *remember that data is a corporate resource.*

The last section in this chapter is aimed at giving special consideration to a problem that can arise regardless of the techniques used for involving the user. This is the resolution of conflicts of information and contradictory descriptions of how the business operates.

The first thing to realise here is that you probably can't resolve these conflicts unaided, so decisions made to resolve the conflict should not be made in isolation. The first form of resolution is to get the appropriate users together around a table and discuss the problem. Obviously, if the user jamboree technique has been used, a number of such conflicts will have been resolved in open forum anyway – one of the advantages of that technique.

If this approach fails to provide an acceptable solution then obviously some other user representative will need to be involved – normally a more senior manager. There are right and wrong ways of doing this. The best approach is actually to get the users to agree that they have a problem that requires a solution and then take this further by actually asking them to recommend the best means of solving it. This ensures their agreement with the means of resolution which of itself increases the chances of such solutions being accepted by all parties.

However, different organisations approach these situations in different ways and it is really up to the data analyst to adopt whichever appears to be the most suitable mechanism. Whatever method is used, two other requirements are important.

Firstly, make sure that the chosen method does work, follow up each referral and ensure that action is being taken in the appropriate manner. If it is not, don't simply ignore the problem as it won't go away. Find a different approach and use that.

Secondly once the conflict has been resolved, do make sure that all interested parties are made aware of the decision and give them the chance to comment. If you keep the decision to yourself and simply slip it into the documentation then you risk losing user confidence. By ensuring all parties are aware of what has happened you are bound to increase user confidence and increase your chances of obtaining full cooperation in the future.

One last golden rule: data is not owned by anyone or by the data processing department; it is a corporate resource, and like any other shared resource there will be some problems over its use. This point should be made to the users and the point must be understood. The data model is not being established for the benefit of only one department but it is being developed for the benefit of the organisation as a whole. It is important to be seen to be independent if the exercise is to succeed.

Chapter 11

Coping With A Complex Model

The Problems

A large data model will give rise to a number of practical problems:

- *difficult to draw;*
- *large amount of documentation;*
- *distribution and control become difficult;*
- *easy to introduce errors over time.*

This chapter will consider some of the special problems that can arise when the area under study is large. The problems are highlighted in this section, and various means of coping with these difficulties are considered in the following sections.

Firstly, of course, the data model can simply become too big to take in. The aim of drawing a data model is primarily that of producing a clear and concise statement of the data structure. There is a practical limit to the number of entity types and relationships that can be conveniently drawn on a single sheet of paper. (For instance, more than 25 entity types on a sheet of A4, or 55 entity types on A3, starts to look very cluttered). One can always resolve this by using larger and larger sheets of paper but the difficulty of copying and distributing the model increases with the paper size.

There will also be problems in actually drawing, and understanding, each relationship as the lines become longer. If it is not possible to see easily where each relationship begins and ends then the model is getting too big. It is also much harder to set up and control the sequence of entity type numbering recommended in Chapter 6.

Secondly, not only does the data model itself become confusing, but the back-up documentation also gets too big to handle conveniently. If a data model has, maybe, 150 entity types identified then there is automatically a large amorphous mass of documentation detailing the entity types and giving examples and volumes. Such a large volume of paper invites difficulties as the model is refined and improved. It is much easier to make mistakes during re-iteration and either to include additional relationships or, more likely, to miss out relationships by accident. Detection of such mistakes will also be more difficult.

Finally, it must be said that however these problems are approached, many of them cannot be solved completely although it is possible to lessen their effect. When all is said and done, if there are 150 entity types, there are 150 entity types! All will need to be defined and detail about each of them will have to be documented.

But enough of the problems .... how do we set about containing them?

"Break it Down"

Do not try to summarise the model but look for ways to split the model either horizontally (by volatility), or vertically (by function). Typical sub-models that can be used are:

- *company infrastructure;*
- *the outside world;*
- *the trading environment;*
- *basic trading;*
- *operations;*
- *accounting.*

The immediate action is to break down the data model into a number of smaller chunks. How should this be done? An approach often adopted by experienced DP staff is to use techniques similar to those used in flowcharting; that is, to draw some form of overview model and follow this up with progressively more detailed data models until the full detail has been presented. There are some major difficulties, however, with this approach.

First of all, it is not easy to summarise a data model without blurring some facets of the model. With a flowchart, which describes a process rather than a structure, the name of a process can be written in the higher-level box and then it is simply a matter of expanding the detail *within the box* at the lower level. This is not possible with a data model. There is no means, nor is it correct, to show a combination of detailed entity types and their relationships as one box on an overview model. It is fair to say that some conventions do support the concept of 'sub-entity types' and these do allow a number of concepts to be shown as one box on a data model. However, the fact that different things are being combined together is intrinsically misleading and, for this reason, this approach is not recommended.

Secondly, one might try to draw a single box to represent several similar entity types. But what should be done about the different relationships? If they are all drawn in there will be considerable clutter. If some are missed out, how is the choice made?

The idea of producing overview models (referred to as 'quick and dirty' models) was introduced in Chapter 6. However, this was put forward as a simple presentation aid only and for use in the initial stages of modelling when there is a need to confirm that the first comprehension of the subject area is moving along the right lines. It is not proposed that 'quick and dirty' models should be used as part of the formal documentation of a detailed data model. They are, almost by definition, misleading and inaccurate. To use them as formal documentation can introduce later inconsistencies.

It is important to never lose sight of one vital principle. *However large the area under study there will be only one data model: to try to present part of it in isolation, would be invalid.*

What has to be done, therefore, is to take the data model and see if it can be split up into several sub-models, each of which would give the full detail for that part of the model, but would also show the interdependency of one model on another. Assuming for the sake of argument that the modelling conventions given in Chapter 6 have been followed and that the whole model has been drawn on a suitably large piece of paper, it should be possible to get some idea about how it can be split up.

First of all, it will be possible to identify parts of the model that are discrete in the functions they serve. These can be divided off and presented as a sub-model. Secondly, by following the conventions there will be a tendency to have the more static parts of the model, representing the 'rules' of the business, at the top of the sheet. The more volatile entity types, representing the actual *doing* of the business, will be towards the bottom of the sheet. Again, the model can be divided along these lines so that the 'rules' are presented as a sub-model and the actual 'doing' of the business on another. This can be seen quite clearly on the data model contained in the Appendix.

This then, is the basic technique. The model can be split *horizontally* by applying the volatility rule which will tend to separate out the static from the volatile. Alternatively it can be split *vertically* which will cause different functions to be separated from each other.

In practice, it is necessary to use a mixture of these two approaches and the techniques take some expertise to master. However, there is a fairly standard breakdown that can be used as a guide to this process. The sub-models which can be used are listed in the appropriate sequence as follows:

1 Company Infrastructure *How the Company is organised, its departments and people.*

2 The Outside World *The way in which the company views things in the outside world including customers, countries, etc.*

3 The Trading Environment *The services or products of the company and their dependency on 1 and 2.*

4 Basic Trading *Details concerning the uptake of products/services by Customers.*

5. Operations *The daily processing of business transactions. "Things that the business actually does".*

6 Accounting *This would include standard financial activities plus any additional entity types required for reporting purposes.*

Depending on the size and complexity of the model, more or fewer categories may be required. If fewer, this is usually achieved by collapsing certain categories together. Where additional categories are needed this can be achieved by imposing vertical divisions within the foregoing breakdown. For example, the rules governing different services or products may be very different and totally distinct so that sectioning them off from each other would add further clarity. The six categories outlined above will apply to almost any environment since each is fundamental to the way any organisation must operate.

There are two important things to remember here. Firstly, the break up is purely arbitrary. The idea is to isolate parts of the total model that can be meaningfully presented together. It does not matter exactly how it is split providing that the resulting models are of a convenient size for presentation in the documentation. This is *not* an attempt to say that each sub-model can be treated as separate from the others, rather that, for the same reason this book is divided into various Parts, Chapters and Sections, it is necessary to break up the whole model into a number of separate, but interrelated chunks.

Secondly, it is important to take some account of the order in which the sub-models are presented. This should ensure that the first time an entity type is shown on a model, then *all* the relationships in which it participates at the 'many' end will be shown. It means that all the entity types upon which the new entity type depends must have already been defined, on the same or an earlier sub-model. This is absolutely essential. It is impossible to define an entity type adequately if all the other entity types on which it depends have not already been defined. Care should be taken to ensure that the proposed splitting of the model will allow this.

This technique implies that some entity types will need to be shown on more than one sub-model. This will need to be done and the conventions for doing it are presented in a later section of this chapter together with examples. The entity type should only be defined once, and this should be on the first sub-model on which it is shown.

As far as the back-up documentation goes, the separate sub-models can be used to break up the entity descriptions. A separate numbering sequence should be used on each sub-model and the descriptions will then refer to the appropriate sub-model and will be presented with that sub-model. This will effectively give a chapter breakdown with each chapter presenting a sub-model and the descriptions of the entity types introduced in that sub-model.

Where this is done, it is also necessary to produce an overall introduction to the whole model, setting out what the data model covers and summarising the structure of the following sub-model. Each should also have its own preamble giving a brief introduction to the main entity types and concepts presented within it.

There are really no hard and fast rules about how the documentation should be produced or the style of its presentation. Much will depend on the standards imposed by the organisation, the size of the data model, and the techniques with which you as author feel most at ease. The main requirement is that the documentation should be presented in a logical and clear manner which will be relatively comprehensible and will not refer to an entity type before it has been defined.

However, it is important to ensure that a consistent approach between the sub-models is retained and the next section of this chapter highlights these points.

Cross Check the Sub-Models

The implications of breaking up the model in a particular way may be quite subtle. Check the following points:

- *have all entity types and relationships been included?*
- *the grouping of entity types within each chapter;*
- *that entity types are defined prior to being referenced by other entity types;*
- *that there is a reasonable 'flow' to the order of the sub-models.*

Having decided the overall structure of the documentation and decided how to break up the data model, it is necessary to make sure that the breakdown is logical and has at least some inherent meaning.

First of all there must be a check that all the entity types and relationships have been included somewhere in the model. This can be achieved *via* a one-for-one check, there is no magic technique here!

Secondly, check that the groupings of entity types within each sub-model or whatever, are logical. Could the flow of the document be improved if some definitions were moved from one chapter to another?

Thirdly, ensure that no entity type has been defined in a sub-model subsequent to where it is first used. Allied to this, check that when an entity type is introduced and defined, all the other entity types upon which it depends have already been defined.

Fourthly, consider the sequence of things that affect each entity type. This becomes of critical importance in relation to process analysis which is covered in Part 3 of this book. For now, though, it will be sufficient to ensure that there is a reasonable flow to the documentation. It is important to 'create' an entity type before it is used.

A very simple example of this is given in Figure 11.1. This particular structure illustrates the sequencing problems that can arise. If it is necessary to present this model across two chapters the most obvious way of presenting this is to define CLIENT and CLIENT JOB in one chapter and the PERSON, TIMESHEET and TIMESHEET LINE entity types in a subsequent chapter.

It is quite likely though, that one would have presented the entity types concerned with the structure of the organisation (including PERSON) at an earlier stage, and the organisation's view of the outside world (including CLIENT) in a later chapter. Assuming this to be the case, how should TIMESHEET and TIMESHEET LINE be handled? TIMESHEET LINE cannot be presented until CLIENT JOB has been defined so this entity type would clearly have to be defined in the later chapter. So where does this leave TIMESHEET? There is little point in defining TIMESHEET in an earlier chapter than TIMESHEET LINE as the two are obviously a logical group that should not be split. So if it is necessary to present PERSON prior to CLIENT for some reason, then TIMESHEET and TIMESHEET LINE should not be defined until CLIENT and CLIENT JOB have been defined.

Obviously with a complex model, the question of exactly in what sequence the entity types should be defined can become quite subtle and may require some thought (as the preceding example should indicate). Care and attention are required if the best sequence is to be established.

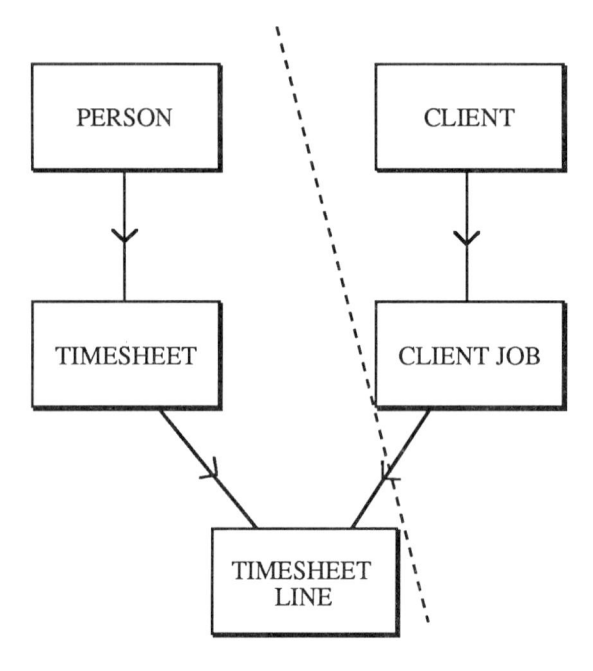

Figure 11.1 Splitting the model along the dotted line is inadvisable

Sub-Modelling Conventions

The following additional conventions should be used in sub-models:

- *where an entity is shown which has been previously defined, indicate this with double horizontal lines on the box;*
- *show a sub-model reference on previously defined entity types rather than the original number;*
- *try to show each relationship on one model, but duplicate relationships if necessary to maintain consistency;*
- *if multiple models (as opposed to sub-models) are used, it is useful to maintain a cross-reference facility in the documentation to show where the same entity type is defined in more than one model.*

This section is concerned with the extra conventions to be adopted where a complex data model has been broken up into several data models.

Figure 11.2 presents the same data model as Figure 11.1 but assumes that at this point, PERSON and CLIENT have already been defined in an earlier sub-model.

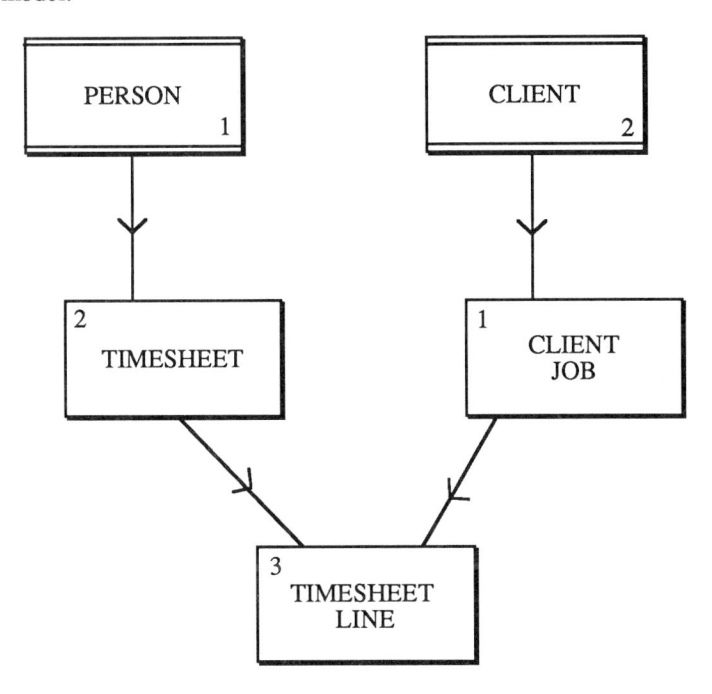

Figure 11.2   A simple Data Model showing pre-defined entity types

Coping With A Complex Model   185

Notice the numbering in each of the entity type boxes. This model assumes that PERSON has been previously defined in sub-model 1. To show this, the horizontal lines at the top and bottom of the entity type box have been added and the reference in the bottom-right hand corner shows in which sub-model the definition of PERSON may be found. The same approach has been used for CLIENT, stating that CLIENT is defined in sub-model 2 of the documentation.

The remaining three entity types are defined in the current sub-model to which Figure 11.2 relates. The sequencing of these is interesting. Both CLIENT JOB and TIMESHEET need to be defined before TIMESHEET LINE. On this example, the decision has been taken to present CLIENT JOB prior to TIMESHEET. This is preferable as it results in TIMESHEET and TIMESHEET LINE following each other and forming a logical group of entity types.

Turning attention to relationships, it needs to be considered how to present these correctly. Generally speaking each relationship should be shown on one sub-model only. The rule adopted here is to show the relationships upon which the entity types being presented in that sub-model depend. As each entity type is defined only once this should ensure that each relationship is only shown once. For example, in Figure 11.2 the relationships upon which PERSON and CLIENT depend are not shown.

The relationships below them, on which the other three entity types depend, are shown. There is, however, one important exception to this. Where the sub-model has a number of predefined entity types shown on it, then any relationships between these entity types should be replicated so that consistency between the models is maintained.

Following these conventions should allow a good measure of clarity between each sub-model and the interrelationships between the different sub-models will be clear, consistent and unambiguous.

However, there are times when the model is so large that it is necessary to take things one stage further. There are occasions when one single data model will not suffice because the subject area is so large that it is necessary to have more than one data model. This is useful where several distinct sets of users are being addressed with little or no overlap, or there are several different data analysts independently modelling different parts of the subject area.

In these situations it is common to discover that, although the areas may be highly independent, many of the entity types will be common. If different data analysts are involved, this may be further complicated by different names being chosen for the same entity types. In all cases, though, it is essential to expend some effort to firstly identify these common entity types and, secondly, to ensure consistency between them. It is essential in these situations that the commonality of the entity types is made clear otherwise design problems may result. Fortunately, there is an easy way of doing this. A Cross-Reference section can be included for each entity description in the back-up documentation. This should identify the other model where the entity type is used and the name used for that entity type in the other data model. This is especially necessary if different names have been used in each model.

Clearly, this is a two-way task and the second definition should be related back to the first. This will require some maintenance, but if properly used will help to ensure consistency between the data models and communication between the data analysts!

Dangers and Their Avoidance

Problems can be minimised via the following:

- *check that the models are consistent within themselves;*
- *check that all relationships/entity types are included;*
- *check attributes carefully;*
- *are entity types shown on more than one model, consistently presented?*
- *do a 'backward-pass' through the examples to ensure consistency;*
- *check for double-definition of an entity type on different models and possibly under different names;*
- *are all entity types at the 'one' end of at least one relationship? If not, why not?*

The final section of this chapter is designed to highlight the special problems that can occur when breaking up a complex data model into several sub-models. Most of these problems have been mentioned earlier in the chapter but this Section is an attempt to pull them all together.

First of all, it is necessary to re-emphasise that the techniques given in Chapter 8 for checking data models during reiteration will be needed even more when several sub-models have been produced. Before even considering the problems of interfacing the sub-models together the analyst should be satisfied that individually they are consistent within themselves.

Assuming this has been done, the first thing to check is that no entity types or relationships have been missed off. With entity types the problem is exactly the same with one model as it is with several and there is no short-cut to painstakingly checking that all entity types are defined somewhere in the data model. However, with relationships the problem is now more complex. Earlier it was suggested that an easy way to check that all relationships are shown is to count all relationships for each entity type and ensure that any differences between one version of the data model and the next are explicable. This is fine where all the relationships for an entity type are shown once and in one place. With many sub-models the relationships which define an entity type (the ones for which it is at the 'many' end) should all be shown where the entity type is first defined and this can be checked, but the relationships which depend on the entity type (where the entity type is at the 'one' end) may be shown in different sub-models. Again care is needed to ensure that all relationships have been shown and that none have been 'moved' or missed out. Checking attributes, especially identifiers, will help to solve this, but there is still a problem.

For entity types which participate in many relationships there is a reasonable way of tackling this. Take the old version of the model and make a note of all the entity types related to the target. Now take the 'new' version and simply compile a similar list. It should now be quite easy to check one off against the other and resolve the differences. This may sound long-winded but it is so easy to miss out relationships and they can be extremely difficult to spot without some methodical approach.

The second thing to check is that a consistent interface has been maintained between the various sub-models. Check that wherever pre-defined entity types have been used on a sub-model, then the structure and relationships between these (if any) is the same on both models. When drawing up a sub-model it is all too easy to forget about the pre-defined entity types as they are not of great concern at the time but these areas must be consciously checked if ambiguity and inconsistency are to be prevented from creeping into the documentation. This also applies to the tables of examples for each entity type. It is well worth doing a 'backward-pass' through the lists of examples to ensure that wherever reference is made to an occurrence of an entity type that has been previously defined, then the occurrence referred to has been listed in the appropriate place. For example, if ACCOUNTS 'belong' to CUSTOMERS then there will be a Customer Number attribute in ACCOUNT. For each example of ACCOUNT that has been given, check that a CUSTOMER with the appropriate Number has been listed in the table of examples for CUSTOMER. This is a soul-destroying task (try it with 200 related entity types!) but it does help others to understand the model if the examples have this consistency.

Another area to beware of is 'double-definition' of an entity type. This can happen where an entity type should be shown as pre-defined on a sub-model but for some reason it has been named differently on both models and, therefore, an attempt is made to redefine it.

This may seem an unlikely situation but it does happen quite frequently, especially where more than one person is involved with producing the documentation.

When revising a data model that is already split into several sub-models it is very easy for some things to fall 'between the cracks'. This is usually the result of attempting to move entity types from one sub-model to either an earlier or later one. The only solution is to adopt a disciplined approach and ensure that whenever you are moving an entity type, a note is made *at the time you decide to move it*, of where you are intending to move it to – but you must *check* that this has been carried through at a later stage.

A final technique for ensuring consistency is to make sure that each entity type is at the 'one' end of at least one relationship. Most entity types will be at the 'one' end of a relationship and, in a sense, this states how the entity type is used. Although an entity type may not be 'used' in this way (especially cross-reference entity types), most entity types will have such usage. Where this does not exist it should be queried as it may be indicative that something is missing or, alternatively, that the data model is incorrectly structured.

We have now covered not only the techniques of data analysis but also how to document a data model and how to gather the basic detail that is needed in order to construct it in the first place. However, examination of data structures is not sufficient for successful system design and it is important to consider a complementary set of techniques which will allow the definition of the other half of the picture. This is process analysis and it follows in Part 3.

Part 3

Process Analysis

Chapter 12

Analysing Processes

Introduction

Process analysis concentrates on breaking down activities into discrete tasks. Two concepts are employed:

- *Operation: a discrete definable activity;*
- *Event: something that triggers an Operation, or is the end result of an Operation.*

Part 3 of this book is concerned with the technique of process analysis and its relationship to data analysis. In basic terms this amounts to a more formalised version of classical, process-oriented, analysis. Data analysis provides one view of the requirements but it is necessary to consider process analysis in some detail otherwise only one side of the equation that leads to successful systems design will have been considered. The importance of process analysis will be discussed in a moment but, first of all ... what is process analysis?

Any successful system does not simply consist of data in isolation. The data that is contained within the system must be used in some way otherwise there is no point in holding it. The usage of this data will be defined by some process, a set of actions that have a particular start point, sequence and end. Process analysis then, attempts to dissect these processes or activities and thus show how data is used by the business. As with data analysis the technique is targetted at the business level. The use or otherwise of a computer system is irrelevant to the subject. If a business is to be effective then the processes will have to be carried out whether or not the business is computerised, and essentially the processes will be the same.

There are two major concepts which must be established in order to master these techniques. These are 'Operation' and secondly, 'Event'.

Operations

Operations are defined as the activities that the business carries out. This is a simple and clean definition and there is no need to worry,

EVENT	OPERATION
Customer books service	
	Booking made and resources allocation
Booking confirmed Customer delivers car	
	Car registration and job sheet completion
Job sheet available	
	Mechanic selects job sheet Carries out work and Indicates this on job sheet
Job sheet returned	
	Garage reception prepares bill
Bill prepared Customer collects car	
	Bill presented and payment obtained
Customer drives off	

Figure 12.1   A high level analysis of the 'service car' activity

for the moment, about any questions concerned with the level at which the operation occurs or its uniqueness within the organisation.

It is central to the application of these techniques that operations are divisible into other, lower level operations, and that these in turn will also be divisible until they have been broken down to the simplest possible form. Taking an example of, say, having a car serviced, the highest level operation might be "Carry out 6,000 Mile Service", the lowest level perhaps, "Remove Oil Filter".

Events

Something else is needed in order to define clearly the start and end points of each operation. There has to be something that 'triggers' the processing (in the above example perhaps 'Customer Brings Car to Garage') and equally there must be something that indicates completion of the operation (perhaps 'Road Test OK'). This is where the second concept, 'Event', comes into play. If an Operation is something that is done, then an Event is something that happens. Events may be external and beyond the control of the organisation, or alternatively, the Event may be the direct result of an Operation.

The interrelationship of these two concepts is fundamental to process analysis. Any activity can be broken down into a series of Events and Operations. Each Operation must be bounded by two Events although it is quite possible for the end Event of one Operation to be the same as the start Event of the next Operation and so on. Figure 12.1 shows the 'Service Car' activity in these terms although the level of analysis is not detailed. As the diagram shows, the activity has been broken down into a number of Operations each bounded by an Event. It is obvious from this example that the Operations can be further sub-divided into more and more detail until a complete and detailed set of indivisible Operations is obtained.

Note that in some cases the End Event of one Operation is the Start Event of the next, and in other cases different. This is not a problem and is quite acceptable as will be seen later.

Notice also that the technique does not at this stage consider the data requirements at all, nor does it assume the existence of any computer system. As with data analysis, the technique is pitched at the business level, using terms and concepts familiar in the business environment.

So why is process analysis important? At a time when database design was less sophisticated, it was thought that a complete and thorough study of the data requirements would be sufficient as a statement for systems design. The argument suggested that if the data was structured correctly then all the processes would automatically be supported in the most efficient way. This has been shown through practice to be a false argument. To design a system that is to be efficient and effective in operation, it is essential to understand not only what data is required but also how it is to be used. A perfect data structure using data analysis may be designed, but this does not, of itself, guarantee that it will be able to satisfy all the usage requirements of that data. It is almost certain that any process to be supported can be handled by the data model but the efficiency of this support may be suspect. Additionally, process analysis may uncover some data requirements which were overlooked during data analysis, so the technique will act as a cross-check on the data model. Later, in the design phase, if it is known that there is a requirement for rapid and frequent access to part of the data it may then be necessary to add additional 'relationships' during systems design to meet this requirement. This information would not be obtained from the data analysis alone but only by also considering the use of that data.

There is obviously a parallel between classical systems analysis and process analysis. As with data analysis, process analysis has always formed a part of successful systems design. The main difference though, is in the formality of the technique. Process analysis does have a specific goal and a clear approach, which should encourage a more thorough and complete result to be obtained. Additionally, by concentrating on a business-level description, it actively prevents the analyst commencing physical system design at too early a stage. There is little point in starting to design an implementation if we have not yet specified what we are designing for!

Operations

When breaking down Operations, bear in mind the following points:

- *is the Operation relevant?*
- *is the Operation purposeful?*
- *avoid ifs, ands, and buts – split the Operation at those points;*
- *split Operations at decision points;*
- *establish the sequence of the Operations;*
- *identify any common or repeating Operations.*

It has been proposed that an Operation is "something the business carries out". Perhaps this should be extended: "for a definite purpose and within the scope of the study". There is little point in considering Operations that are outside of the terms of reference or have no bearing on the subject area, but even here some care needs to be exercised. It is a mistake to exclude Operations which are important but which are at first sight irrelevant, therefore there must be a definite reason for excluding an Operation. Some of these Operations may not be the responsibility of the main user department so it is worthwhile talking to other interested departments to ensure that their requirements are met.

Equally, though, an Operation must be purposeful: it must achieve something. If this cannot be established for an Operation then its relevance should be questioned. As part of this exercise there should be an attempt to describe the Operation. What the Operation entails should be stated clearly and at this stage, this should be done at the business level. The interplay between the data model and the Operation will be considered later. The aim at the moment is to address the same problems from a different and independent view point. In beginning to describe each Operation, the natural breakpoints will start to emerge and it will be possible to proceed to greater levels of detail. But how far should this be taken? There is no cut and dried answer to this. Primarily it is necessary to isolate the essential 'building blocks' of the process. Each Operation should be broken down until there is a definite statement of what the Operation does, with as few uncertainties as possible.

If a decision needs to be made and a choice of paths is made as a result, then the Operation should be split so that there is one Operation ending at the decision point and then a further Operation for each path from the decision.

As can be seen, there will be a sequence to the Operations and the establishment of this sequence is important. Additionally, some Operations will be common to many 'higher level' Operations or may occur several times. It is important to establish these Repeating Operations and Common Operations as they will be of importance in later design.

Events

It is possible to define two types of event:

* *External: over which the organisation has no control;*
* *Internal: marking the boundaries between operations and indicating their completion.*

What is meant by Events in this context? This book uses an empirical definition, simply stating that an Event either initiates an Operation or that an Operation has an Event as its end result, or both. It is of importance that one will tend to talk of Event in a passive sense (for example 'Claim Received'). In other words, nothing is done, the Event 'happens' of its own accord. Operations tend to be active (for example 'Process Claim') implying that something has to be done. It is this passive/active dichotomy that is fundamental in distinguishing between Events and Operations.

However, it is useful to step back a little and contrast Event in this context, with the concept of Event as used in a pure data modelling sense. In the data modelling sense, Events are only of importance if there is a need to record the fact that they have happened. In other words – is there some data or information that needs to be recorded as a direct result of the Event? But in the process analysis sense, the concept of Event is much wider. It is certainly necessary to consider all the 'Events' that were uncovered during data analysis, but equally certainly, there will be a need to establish many others.

Process analysis Events will fall into two groups and it is useful to distinguish between them. Firstly, there will be those Events that are external. These are things over which the organisation has no control but which nonetheless will initiate a sequence of Operations. The Event 'Claim Received' given above is a typical example. The Operation is triggered as a direct result of the Event and will be controlled by the organisation, but the organisation cannot control how many Events of this type will occur or when they will occur.

It is also of note that External Events are normally at the start of a string of Operations. It is usually, in fact almost certainly, incorrect to have External Events in the middle or at the end of a sequence of Operations.

The majority of Events, though, will be internally defined and created. They will mark the boundaries of the tasks that the organisation carries out. Internal Events may occur at the start of a sequence of Operations or at the end or anywhere in the middle. It is also true that many of these Events may seem somewhat artificial either because they mark operational boundaries in the organisation (that is, who may do what) or because they have been invented to break up larger more complex processes into more convenient portions. This is quite acceptable. Such Events have normally been introduced for sound business reasons and, unless there is a good case for change, these Events should be included in the analysis.

Levels of Nesting

> *To arrive at the 'right' level, note these points:*
> - *break down the activity into the obvious main steps;*
> - *continue this until there is a list of non-divisible activities;*
> - *check that the list of operations when recombined covers the whole of the original activity;*
> - *establish the sequences;*
> - *check for any further embedded decisions and break these out;*
> - *identify any repeating or common Operations;*
> - *identify any 'time delays' and break these out.*

One of the main difficulties with process analysis is in knowing how far each Operation should be decomposed and how many intermediate Events need to be interposed along the way. This is something which can only be learnt from experience but there are a number of guidelines that can be explored.

First, take the concept of 'divisibility'. It is relatively easy to produce a short-list of high-level Operations for the organisation. This will be little more than a description of the main activities undertaken by the business. Having produced this, it is necessary to determine the constituent parts of each of these Operations. This need not be exhaustive or necessarily correct at this stage, but rather it is a first-cut at decomposing the high-level Operations. This is effectively what was shown in Figure 12.1 for the 'Service Car' Operation. Having done this, it is possible to ask if each of the lower-level Operations is meaningful in itself; can it be carried out in isolation, or does the Operation only make sense in the context of preceding or following Operations? Where the Operation does not make sense by itself it should be recombined until there is a list of Operations each of which makes sense by itself, yet which, when all are combined in a given sequence, constitute the high-level Operation initially chosen. This exercise should then be repeated at the lower level and a further level of Operations identified. This process can be continued until there is a list of Operations each of which is non-divisible and cannot meaningfully be divided further.

This is now starting to get the analysis well along the way and there is little point in breaking down the Operations any further except for two important situations. The first of these concerns decision points within the Operations so far defined.

Let us be perfectly clear about this. If there is a decision at the end of an Operation to determine which other Operations to initiate, there is no problem. What is of concern here are decisions and choices which are still embedded within an Operation. This may well have occurred because Operations have been combined together that would not stand up on their own. Unfortunately, now is the time to split them out again. Wherever there is such a choice of routes through the list of Operations, an Event must be interposed and the Operation split down. As will be seen later, there is a need to be able to define clearly the data each Operation will use, so if there is a decision to be made, there is no longer any certainty as to what the data usage will be.

Lastly, look for common Operations, or common elements within some of the Operations. If an Operation is recognised as occurring in several places then this should be noted to ensure that it is only defined once. More importantly, it may be possible to define elements of some Operations that are used as part of some other. Again, these should be split out to separate the common processing into an Operation of its own. This is analogous to standard sub-routining and its aim is the same. There is no point in defining the same thing twice, rather it should be defined once and used in as many places as necessary.

Let us take a sample application and illustrate the rules with some examples. Figure 12.2 suggests a list of main business activities or Operations for a motor insurance application. This is not a full list but it is sufficient to give a feel for the level of activity required. Figure 12.3 details a first-cut list of Events and Operations within the Incept New Policy activity. This is deliberately rough-and-ready but does distinguish the main Operations within 'Incept New Policy': these are 'Inputting of Policy Details', 'Premium Reconciliation', and 'Issue Policy Documents'.

This gives a start point from which the Operations can be broken down more fully. The results of this are shown in Figure 12.4. Notice that the decision points embedded in the higher Operations have been split out into separate Operations and the start and end Events have been isolated. As a result of this there is no longer a fixed series of Operations, but rather a number of Operations which can be strung together in a number of sequences depending on the decisions made. Furthermore, some of the Operations, such as 'Diarise Follow-up', appear more than once and should also be separated. Figure 12.5 lists all the intermediate Operations resulting from this and also the permitted sequencing that can occur.

Figure 12.6 takes this a little further by showing, using simple flowcharting techniques, the breakdown of the Process into the specific Operations of which it consists. Such a flowchart depicts the permissible sequencing of Operations and makes the decision points explicit. One other advantage of this technique is that time-delays or other breaks in the process flow can be clearly depicted. For example, checking follow-up detail received must follow the diarising of the follow-up but there is an important time-delay that needs to be reflected. This is shown in Figure 12.6 by the use

```
            MOTOR INSURANCE APPLICATION

            1     Give Quotation

            2     Incept New Policy

            3     Adjust Existing Policy

            4     Renew Policy

            5     Enquire on Policy
```

Figure 12.2   Main Activities for a typical application

```
EVENT                    OPERATION

Proposal received

                         Input details and calculate premium


Base detail created

                         Match premium calculated to premium
                         received and take action on difference
                         (if any)

Premium detail created

                         Issue policy documentation and post to
                         Underwriter

Policy incepted
```

Figure 12.3   'First-cut' Operations and Events
                   for 'Incept New Policy' activity

of an open-sided box. Further examples of these charts are given in the Appendix.

Two other conventions are required which are not shown in Figure 12.6. Firstly, it is possible for an Operation to 'fire-off' two parallel streams of processing without any decision being needed. This can be depicted, quite simply, as two lines emerging from the Operation as shown in Figure 12.7. In this example, the process "Check Claim Form" results in two processes being initiated, these being "Appoint assessors" and "Contact other parties involved". These latter two processes are independent of each other and may occur simultaneously.

Secondly, the converse of this may be required: that is, an Operation may only begin when two or more preceding Operations have been completed. This is shown in Figure 12.8. In this example, 'Arrange Payment' can only begin once 'Approve Claim' and 'Receive Receipted Account from Garage', which are independent, have *both* occurred. If it was that *either* 'Approve Claim' or 'Receive Receipted Account From Garage' could initiate 'Arrange Payment', then this would be shown by joining the lines together *before* the box, as per Figure 12.6.

This illustrates the basic aim of process analysis – to take a high-level activity and break it down into its constituent parts. The examples above illustrate the main points but there is a further aspect not shown above. Figure 12.6 considers only the 'Incept New Policy' Operation. When the other Operations in Figure 12.2 are considered, such as 'Adjust Existing Policy', some of the lower-level Operations will be common between the two and this will need to be recognised. 'Issue Refund' would be a good example of this. If these are the same process, and they probably are, then this must be recognised.

Finally, a word of warning. There are dangers in taking process analysis too far. Taken to its logical extreme, the end result is virtually program code written in English with every nuance of the logic spelled out. The aim is not to specify programs at this stage! It is important to be conscious of the guidelines above. Is the Operation divisible? Is there a major decision embedded in the Operation? Is there a significant portion which is common to other Operations or is repeated? If the answer to all three questions is 'no', then there is no need to break the Operation down any further. Examples of this documentation can be found in the Appendix.

START EVENT	OPERATION	END EVENT
1 Proposal received	Input static data (Customer and address)	Customer detail created
2 Customer detail created	Input risk data and check complete	Risk data created
3 If risk data missing	Request further data Issue cover note Diarise follow up	Policy in 'suspense'
4 If risk data O.K.	Calculate premium and check with premium received	Premium due calculated
5 Premium due equals received	Issue policy documents and post to Underwriter	Policy incepted
6 Premium due greater than received	Issue policy documents, post to Underwriter and arrange refund	Policy incepted
7 Premium due less than received	Issue cover note Issue payment request Diarise follow-up	Policy in 'suspense'
8 Diary date prompt	Check additional detail received and either recalculate premium or diarise new follow-up	Diary prompt processed

Figure 12.4 Detailed Operations for 'Incept New Policy' Activity

1   Input Static Data

2   Input risk data and check complete

3   Request risk clarification

4   Issue cover note

5   Diarise follow-up

6   Calculate premium and check with premium received

7   Issue policy documents and post to Underwriter

8   Arrange premium refund

9   Issue payment request

10 Check follow-up detail received

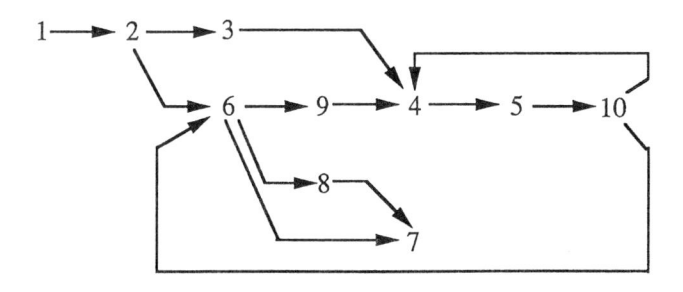

```
1 ──▶ 2 ──▶ 3 ─────────────┐
        ↘                   ▼
          6 ──▶ 9 ──▶ 4 ──▶ 5 ──▶ 10
        ↗ ↘
        │  ▶ 8 ─┐
        │       ▼
        └────── 7
```

Figure 12.5   List of Operations and Permissible Sequences

Analysing Processes

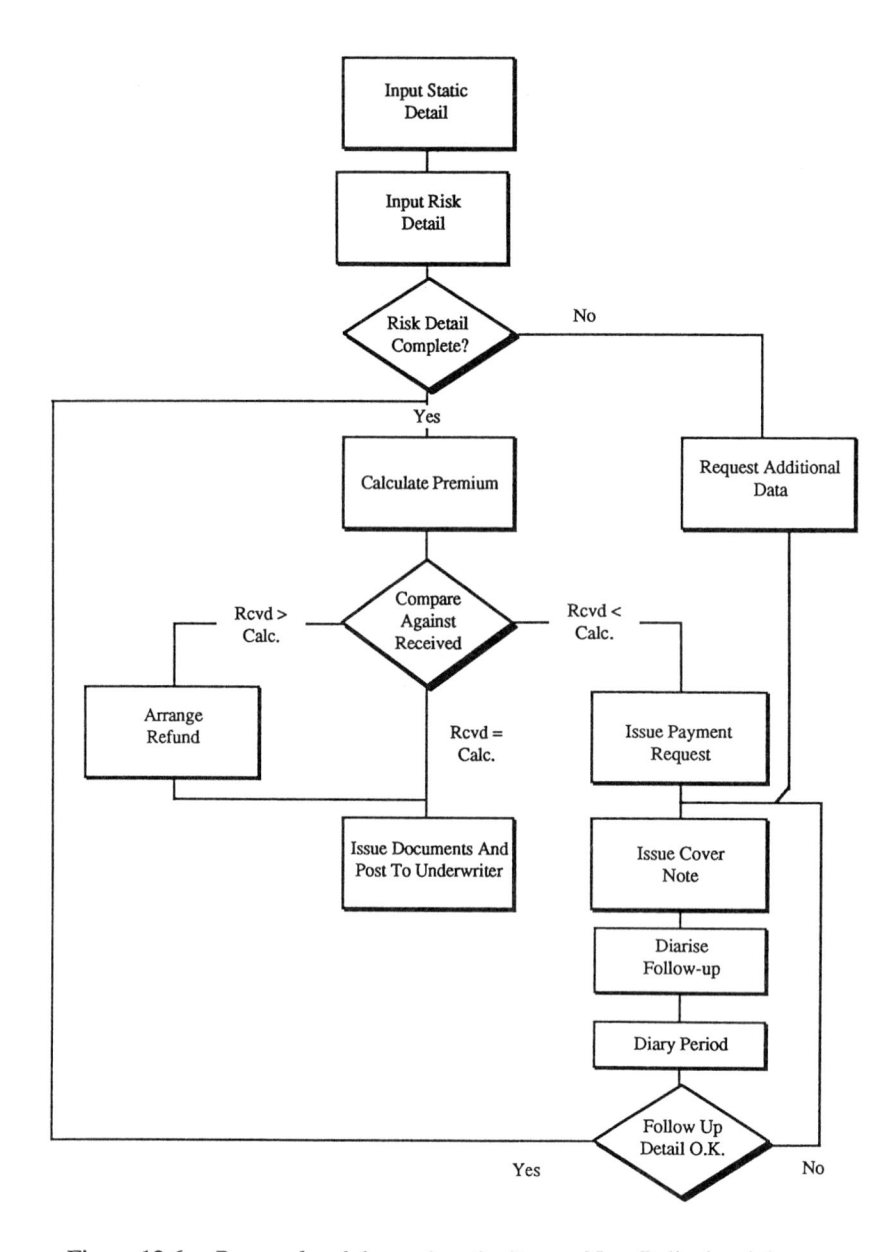

Figure 12.6   Process breakdown chart for 'Incept New Policy' activity

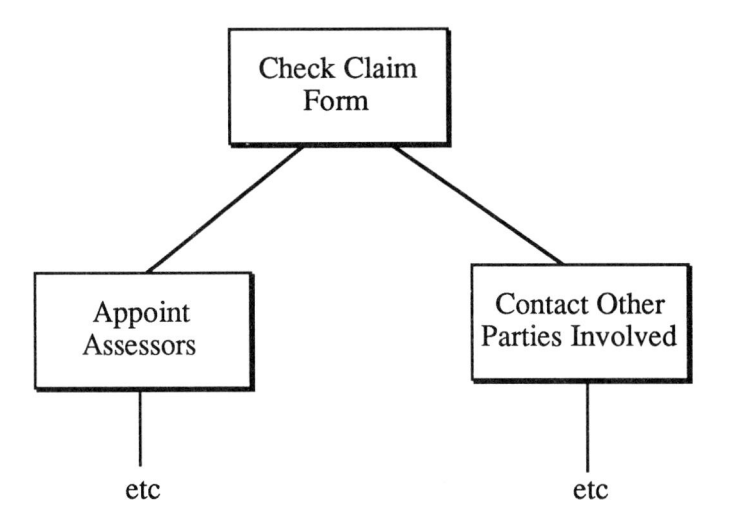

```
              ┌──────────────┐
              │ Check Claim  │
              │    Form      │
              └──────────────┘
             /                \
  ┌──────────────┐      ┌──────────────────┐
  │   Appoint    │      │  Contact Other   │
  │  Assessors   │      │ Parties Involved │
  └──────────────┘      └──────────────────┘
         │                      │
        etc                    etc
```

Figure 12.7   Initiation of two streams of parallel Operations

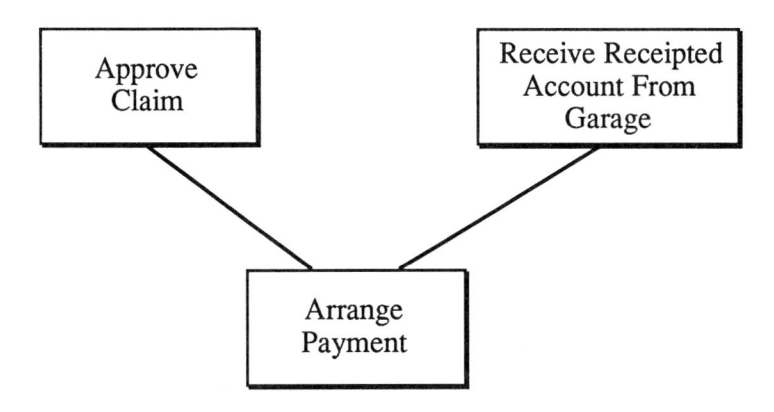

```
  ┌──────────┐            ┌──────────────────┐
  │ Approve  │            │ Receive Receipted│
  │  Claim   │            │   Account From   │
  │          │            │      Garage      │
  └──────────┘            └──────────────────┘
          \                    /
           ┌──────────────┐
           │   Arrange    │
           │   Payment    │
           └──────────────┘
```

Figure 12.8   Showing an Operation as dependent on two others

Other Useful Data

In addition to breaking down the Operations, further information should be gathered and documented:

- *describe the Operation in English;*
- *who is responsible for the Operation?*
- *who carries out the Operation?*
- *establish events for start and end;*
- *how often is the Operation carried out?*
- *what is the response time?*
- *what is the criticality?*

The foregoing sections detailed the basics and the aim of process analysis, but a simple list and sequence of Operations is insufficient for this purpose. Further information is needed and it is most important to gather *and document* this detail.

Firstly, describe the Operation in simple English. An Operation such as 'Diarise Follow-up' must be described fully to ensure that others fully appreciate what is meant. The length of the description will vary with the complexity of the Operation but normally a few sentences, maybe 100-250 words, will suffice. This will often lead to a further realisation of either major decision points or common processes and will, therefore, help to refine the Operations list and sequencing.

Secondly, establish who carries out the Operation and how often. Establishing the frequency is fairly straightforward and, as with data analysis, an order of magnitude is needed rather than a precise figure. The 'who does it' question is more complex. There are two options here and both should be considered. First establish who is *responsible* for the Operation – which parts of the organisation have the job of controlling and managing the process. Secondly, consider who actually *does it*. How many people are involved? Is the Operation carried out in several places depending on some other factor?

Next check that both the Events that trigger the Operation and the Events that signal its completion have been established. Establishment of the End Event effectively describes the end-product of the Operation and, as discussed above, this Event may itself trigger other Operations.

Fourthly, the frequency gives the remaining information required in order to design an eventual implementation. It is now known what each part of the organisation does, how often, and the necessary sequencing of these Operations, but a further piece of information is required.

The criticality and response time required for the Operation must be established. How quickly is a response required during normal day-to-day operation? This will often be interrelated with the frequency of the Operation. If the Operation takes place 300 times a day, for example, then it is quite likely that a rapid response will be required. Where an Operation happens once a month, a much longer response time may be acceptable. Nevertheless, some feel of the speed for response must be gathered.

Criticality is a more nebulous concept. This tries to establish, in relative terms, the importance of each Operation. How critical is the Operation to the day-to-day running of the business? What would be the effects of failure or poor service in this area? Obviously, all Operations are important but a relative judgement can be made. In retail banking, for example it is obviously more critical to be able to obtain the current balance on an account than it is to produce a list, say, of all the standing orders active on an account. A judgement of this kind will again feed into systems design to direct resource to those parts of the implemented system that are the most critical, most frequent, and require the quickest response. How this detail is used to refine the design is considered later.

To round off this chapter, it is necessary to discuss where this information comes from. The data analyst cannot gather this detail unaided and the user departments will have considerable input to this process. Once again, it is the users who are the repository of most of the information that is required. As with data analysis, the detail can be obtained from many of the same sources and it is useful to undertake process and data analysis together so that each source of information is only consulted once.

This is preferable to running two separate sets of meetings one concentrating on data, the other on processes. Both processes and data are highly interrelated anyway so it can be very useful to consider both at the same time.

Chapter 13

General Techniques for Process Analysis

Types of Process

There are two independent classifications of processes:

* *Purpose*
  * *– Day-to-Day Operations;*
  * *– Imposed Internal;*
  * *– Imposed External;*
* *Nature*
  * *– Updating Processes;*
  * *– Regular Enquiries;*
  * *– Ad-hoc Enquiries.*

By looking at the techniques of process analysis in Chapter 12, the aims and purpose of this approach have been explored. This chapter concentrates on a number of points that apply to process analysis in general and as such need to be borne in mind by the analyst.

There are two useful classifications of processes. The first of these concerns the purpose underlying the process and has three categories. These are: Day-to-Day Processes, basic functions to achieve the objectives of the business; other processes are Imposed either because this is the way the business wishes to regulate itself, or because of some external requirement. The second classification considers the nature of the process and is independent of the first classification. Again there are three categories. First of all, Updating Processes, which create data in the database. Secondly there are Regular Enquiry Processes, which do not add any new data, but are necessary to report on and to monitor the results of the first category. Thirdly, there are Ad-hoc Enquiries, the most difficult category to estimate and control but nonetheless a very important grouping.

It should be remembered that although these categorisations are independent, they do overlap to some extent. Figure 13.1 shows some typical examples.

PURPOSE / NATURE	Day-to-Day Processes	Self-Imposed Processes	External Processes
Updating Proccess	Incept New Policy Renew Policy Adjust Policy	Write-off Bad Debts	
Regular Enquiries	Display Policy Details Correspondence Enquiruy	Outstanding Claims Report Management Reports	Dept-of Trade Return Tax Return
Ad-Hoc Enquires		"Have we any Customers called 'Smith?" "How many Fords are insured?"	

Figure 13.1   A process breakdown matrix for the 'Motor Insurance' application

The Purpose of the Process

Each type of Process has its own features.

- *Day-to-Day Processes*
  - *high frequency;*
  - *efficient processing;*
  - *must be effectively supported;*
  - *the actual and declared process may vary;*
  - *gather accurate statistics about them.*
- *Self-Imposed Processes*
  - *primarily extractive in nature;*
  - *used to control and direct;*
  - *may reflect artificial boundaries;*
  - *scope – do they apply to the whole organisation or just a part?*
- *External Processes*
  - *as for self-imposed processes, plus*
  - *the organisation has no direct control over them;*
  - *they must be adequately supported.*

Day-to-Day Processes

When considering the purpose of the process, Day-to-Day Processes are by far the easiest to isolate and define. These processes support the basic functions of the business. The best people to provide the required detail are end-users, as they are more aware of what actually goes on at the sharp end.

Although these may be the easiest processes to isolate, they do have their own peculiarities that should be recognised. First of all, Day-to-Day Processes will often be performed frequently and it will be necessary to design a well-regulated and efficient means of carrying them out. This places a great responsibility on the data analyst to ensure that these processes have been understood and that the requirements of each have been fully realised. It is important to understand not only what happens inside each process, but also the exact interrelationships between them. The permissible sequence of operations will be highly important. The initiation of one operation by another is something which will have to be built into the eventual implementation. If these subtle controls are missed at this stage, the result will tend to be a very 'loose' system's design which will cause much difficulty later on.

Secondly, the actual and the declared process may well be different. Although it may be relatively easy to pick up the theoretical description of the process, it will be important to make sure that this is indeed how the process is carried out in practice. It is human nature to try to find short-cuts and easier ways of doing things and this is more likely to occur where a process is frequently performed. Wherever such a dichotomy becomes apparent, corrective action should be taken and consideration given to the best way of doing things. Is the short-cut of real benefit? Does the short-cut lead to potential control difficulties if problems should arise? The available choices will have to be discussed with the relevant parties.

It is possible for the analyst to hit on a different way of carrying out a process that may be easier, more efficient, or have certain benefits over the existing method. Once more, these choices should be carefully discussed. It is very dangerous to assume, just because it is your idea and you have thought about it long and hard, that you have automatically come across a better way of doing things.

Lastly, care must be taken over the statistical information gathered about Day-to-Day Processes. It is important to get accurate data about exactly how often the process occurs and how quickly it must be carried out. A serious miscalculation in this area could invalidate the whole of the eventual implementation, so beware ...... check this data thoroughly!!

Self-Imposed Processes

The second category of processes that has to be considered is that which I have termed Self-Imposed Processes. These are not central to the day-to-day running of the business but are employed to control and direct the business. In this sense, these are the processes that support the view that the organisation has of itself and its place in the outside world.

It is very easy to fail to recognise the importance of these processes. Many of them will not be performed particularly frequently and, depending on who is consulted, they may not be stressed in discussion. The aim is to obtain this information from middle and higher management. All businesses have a need to monitor and control their everyday operations and it is these processes that give the key to what the requirements are in this area. These processes will be primarily extractive, that is, they will not so much create new data, as collate and analyse data that is already available. Beware, if there is a structure that requires supporting, then there must be some processes concerned with maintaining and controlling this structure.

Failure to take adequate notice of these things at this stage could make it extremely difficult to supply management with the right information at the right time, and in the right way.

However, this should not be restricted to purely internal processes as the organisation will also impose its own view of the outside world which may well be different from reality. This will also need to be supported. These artificial views are important and it is necessary to question each of them to see if they should be perpetuated. If there is a genuine business reason to justify why a particular structure is being used then there is no need to find an alternative approach. Many of these artificial views may well have been imposed because of some restriction in the way the current systems or procedures work. It may be that because of some other problem, perhaps in the inflexibility of the current system, the business is forced to take a particular view. These can be difficult to uncover. If a system has been running successfully for some time, the artificial constraint may have become very ingrained and the users of the system may accept the problem as a perfectly natural factor. Wherever possible these artificial constraints should be uncovered and agreement obtained to remove them. Once again, do not simply discard them without thought. Check your facts and obtain full agreement before hand.

Lastly, it is necessary to establish the scope of these imposed processes. Do they apply across the whole organisation or only to certain parts of it? If the imposition is not organisation-wide, how is the interplay between different sections reconciled? Are there two or more different views used by different sections? The analyst has to tread very carefully here. What is an almost unworkable constraint on one part of the business may be absolutely crucial to the efficient operation of another section. It will be necessary to assess the relative merits of each, and if possible, bring the two together.

External Processes

External Processes refer to those activities that need to be carried out at the request of, or by the order of, some external body or organisation. As with self-imposed controls, these processes are primarily extractive and it is relatively unusual for these to result in the creation of data. Examples of this type of process would be tax returns, government statistics and reporting, and so on.

The important thing about these processes is that the organisation under study has no control over them. If these processes are to be supported in a computer system then this has to be flexible enough to deal with changed requirements quickly and easily. This can be extremely difficult. There is no way of predicting when changes are likely to occur, what the changes will involve, and how much time will be available to respond to them.

It is essential that the data structures used to support these processes are built in a very flexible way and that as many of the imposed 'rules' as possible are explicitly defined as entity types within the data model itself. In this way, it is possible to parameterise the process to some extent, which will assist us when changes occur. These processes are highly important: failure to carry them out accurately and on time can result in the business simply being closed down – and we do not want to be the cause of such events!

It is essential that adequate time is allowed to study and understand these requirements. It is necessary to discuss them at length with the user departments that have to carry them out. The users will often have some feel for how volatile the requirements are, and how strict the recipient will be. This information is crucial and it is imperative that the importance and complexity of these processes is not underestimated.

The Nature of the Process

There are specific features to each of these categories:

- *Updating Processes*
    - *sequence;*
    - *Who? When?*
    - *time-criticality;*
    - *error correction;*
    - *internal audit requirements.*
- *Regular Enquiries*
    - *do not underestimate;*
    - *analyse to understand;*
    - *collect statistical data for later design.*
- *Ad-hoc Enquiries*
    - *difficult to estimate;*
    - *be generous;*
    - *classify as much as possible.*

This section discusses the second of the two classifications or types of process and hinges on the way in which data is used by a process. There are three categories: Updating Processes, Regular Enquiry Processes, and Ad-Hoc Enquiries.

Updating Processes

Updating Processes are of vital importance to system design. All data has to be created, and possibly updated at some time, and it is imperative that the processes that achieve this are correctly understood and assessed. The sequencing of these processes is critical. The creation and update of data cannot be treated in isolation within an organised and structured collection of data items. It is necessary to be very clear about what data must exist *before* the process is invoked, and also about the effects of the process on that data.

Updating Processes are often highly time-critical. It is frequently very important for the data to be recorded in the system quickly and easily and these requirements should be studied carefully. Should the new data be available for enquiry immediately after creation? What is the response time required for the process? Allied to this, the volume of transactions may be relatively large. If there will be many updating processes running together, how should they be interfaced? Will they interfere with each other? Additionally, a process may be invoked by several different parts of the organisation. This may cause problems, not only in determining who should have the authority to carry out the process, but also how this is to be controlled from an organisation point of view. Wherever possible each process should be carried out in one user area only.

If there appears to be a multiplicity of units that action it, the process should be examined to be sure that it is the same in all areas, and to investigate the possibility of bringing them together in one place. Where this cannot be achieved easily, then examine the control of the process to ensure that it is adequate.

There is another unique feature of updating processes which requires special note. If there is an update process then there must almost certainly be an 'un-updating' equivalent, that is, some process that can be invoked to correct or back-out the effects of errors or other mistakes.

The tightest solution to this problem is simply not to allow data once created to be changed, but to allow reversing entries to be made so that both the original error and its correction are equally visible to later enquiry. This is standard accounting practice and in this instance is highly recommended, though it does require more planning. If this is to be done effectively, all processes have to be 'aware' that errors and corrections may be encountered and they have to be designed to handle these situations correctly. If an incorrect entry is made on an account and then reversed out, for example, does the organisation wish this to be shown on any ensuing statements, or does it prefer to net out the two entries thus avoiding the embarrassment of having to own up to its own mistakes?

The data analyst alone cannot answer this question. The degree of control and audit trail required is a valid question for user consideration. In some cases of course, this is laid down by either legal constraint or on the advice of some other external body (for example the organisation's auditors). These questions are very important and must be fully discussed.

This is where advice from the relevant accounting and internal audit department should be sought. It is all too tempting to treat internal audit and accounts as interfering departments that are better ignored but, if you seek out their advice at an early stage and ensure their requirements are met, you will save many problems later when they eventually tumble to your project and arrive unbidden with their requirements!

Lastly, it is easy to assume that these kinds of controls apply only to accounting or other financial transactions. This is simply not true. It will be necessary to take the same approach to *all* areas where data is updated if the eventual system is to be correctly designed and controlled.

Regular Enquiry

It is often stated in discussions on process analysis that if the analyst takes all due regard of updating processes, and establishes a good data structure, then the reporting requirements will look after themselves. Unfortunately, life just is not so straightforward. Anyone with experience in a large, possibly bureaucratic, organisation will be aware that the number of reports and other summaries required can make significant inroads into the data processing budget and machine resources. To make matters worse, with the ever-broadening use of computers, these reporting requirements are likely to increase as the volume of computerised data increases.

One of the good things about regular reporting is that it is quantifiable and in many cases, relatively straightforward. To start with it is usually fairly easy to identify the processes and obtain clear unambiguous data on how often they occur. So far so good! However, the main difficulty arises in trying to establish just how the report is compiled.

This is not too bad if the process is imposed by the organisation itself as there should be someone, somewhere who understands the requirements. With some externally imposed requirements, though, the regulations governing how these are to be compiled can be very daunting. Once again, the analyst must go cap-in-hand to the users and request their help. It should be possible to unravel the details by working together on the problem.

A clear understanding of these processes is essential fodder to later design efforts. By taking no account of how these requirements are satisfied, it could turn out that twelve hours of processing have to fit into an eight hour night or, worse still, that there is a lengthy on-line process that brings the entire system to its knees.

Ad-Hoc Enquiries

The third category, Ad-Hoc Enquiries, is the most difficult of all to analyse! As the amount of computerised data increases, so users want to interrogate that data more frequently, often in different and unpredictable ways. Added to this, the very nature of this type of process tends to mean that the enquiry is ill-defined and the user may make several similar enquiries before he has honed the question to the precise format required. An Ad-Hoc Enquiry, by definition, is unpredictable and to a great extent un-quantifiable.

Nevertheless one must try to deal with this difficult situation and come to terms with it as much as possible. The analyst must ask the users (yet again!) what sort of enquiries they might want to make and what data will be searched. It is important to develop a 'feel' for this type of usage. Are the users sufficiently aware and motivated that there will be high usage or do they want to ignore the computer as much as possible? How much resource does management want to commit to servicing these enquiries? How much can Ad-Hoc Enquiries be allowed to degrade the on-line service?

The answers to these questions are not likely to be very clear-cut. One thing is certain: if a successful design is implemented then the volume of such enquiries will increase beyond predictions! As we all want our systems to succeed, be generous in this area. Allow a good percentage over the expected volumes and try, where possible, to quantify just what volume of

Ad-Hoc Enquiries can be handled before there is a noticeable degradation of existing services. At this stage, do *not* consider any solutions to the problem. How the predicted workload is serviced will be considered much later. At this point all efforts should be restricted to the basic analytical skills of trying to quantify what will be done, how often, and by whom.

People and Positions

When considering the responsibility for authorising, controlling and actioning a process, bear in mind the division between person and post. Use the level appropriate for the organisation and compare with the approach used in the data analysis.

The next point to be considered is very similar to a standard data analysis problem except that it is viewed from a different angle. This is the person/post dilemma.

The statement that "Joe authorises loan requests" is ambiguous and the control and management of the process will depend on what is actually meant by the statement. Will Joe continue to authorise the same loans if he is promoted, or is this the result of the post he currently occupies? The severity of this problem will depend on the formality of the organisation under study. For many large, highly structured organisations, the concept of 'post' is readily distinguishable from that of 'person'. In the event of staff turnover or promotion, the post is filled by a different person but the responsibilities and duties of the post remain unchanged.

This is obviously relatively easy to model from both the data and process viewpoints. In many smaller or more specialised organisations the loss of a single member of staff could result in a complete re-organisation of the department. In these situations the concepts of person and post become synonymous in the mind of the users. Employees are not simply 'slotted in' to a suitable vacancy, rather the job is split up over the mix of skills available across the people in the department. In this case the analyst will need to dig down to discover how useful the separation of people and posts might be. If the two can be separated then the result is usually cleaner and more flexible but, on the other hand, the concept of 'post' may be so alien to the user that to force this solution would simply cause more problems that it solves.

Implementation Constraints

When studying processes, try to detect the following situations:

* *process breaks made for administrative convenience;*
* *artificial constraints imposed by an existing system;*
* *beware old habits which may have no sound justification.*

This section discusses constraints that are due to the limitation of existing systems, whether these be manual or computerised.

Let us examine the manual processes first. There may be some doubt as to what this means. A manual process must be a valid process to consider because the business would not carry out the process if it were not necessary! This is almost certainly correct: the process probably will be necessary, but it is the purpose of the process that is of concern here. The situation being considered is not common but does occur. It happens when it is necessary to manipulate some data purely for the purpose of making some other process easier. Typically this might be copying selected data to another file so that this may then be used for some purpose, normally managerial or statistical. In an ideal world this should all be one process – extract the data and manipulate it – rather than two. The split between the two is artificial and is made for administrative convenience. When documenting the processes such artificial situations should not be carried over. This returns to the point, made at the beginning of Part 3, that a process must have a clearly defined purpose and end result. Difficulty in establishing these factors may be caused by a manually imposed constraint which is not fundamental.

By far the most common form of implementation constraint, though, is due to the design and function of existing computer systems. These constraints will occur where the functionality of the system is different from the user's needs or is lacking in capability. This sort of problem can be detected in several ways.

You may be very lucky and be directly informed of the problem either by the users or DP personnel. In other cases the clues will be more subtle. When probing the details of a process, the analyst may start to encounter phrases such as "because the computer does it that way" or its equivalent. This is a reliable sign that something is artificial and it will be necessary to uncover the real underlying functionality required. Once again the new design should not be predicated on old difficulties.

There is a more extreme version of implementation constraint that occurs where systems have been well established for a long period. The analyst may find that users are not able to give any explanation for why a process is done the way it is. The reasons are lost in time and the custom has become ingrained over years of repetition. This is very dangerous: because this may well be embarrassing for users to admit, there will be a tendency to steer away from the problem and either keep it as it is or, even worse, ignore it altogether. Neither course is acceptable. It is important to understand each and every process and evaluate its importance. If a process is not understood the evaluation of its usefulness cannot be meaningfully made. Therefore these processes must be investigated with care to establish the requirements.

Historical Processes

There are two basic things to consider here.

- *data is rarely needed 'for ever': what is the ideal retention period and how is it controlled?*
- *there may be processes designed to control future events; these should also be analysed.*

The importance of historical data has been discussed at length earlier in this book, but it is of no use at all if there is no process to manipulate it. These processes must be established. These may be comparative processes, examining current and past situations for some purpose. Each will need to be analysed as a guide to how long data needs to be retained. Year-to-date or similar summary information will be important here. There are few businesses that do not use historical data in some way. Part of the analyst's task is to discover just what these processes are and how historical data is used.

Allied to this, yet separate, is the process of discarding old data. Many organisations will have a means of weeding-out data they no longer require. The conditions that need to be satisfied before this process can be initiated should be established and the process examined very closely. But caution is needed here. Many of these processes will be controlled by the amount of data the organisation can efficiently handle with its current systems, be they manual or computerised. There may be problems as a result of this and the users may prefer to have information retained for a longer period if possible. Be practical about it. If the organisation keeps data for twelve months and they are asked if they would prefer to retain it for five years, the answer will almost certainly be 'yes'. Do attempt to discover if the twelve month cut-off actually causes any problems – and if so how frequently. Use this information to determine the ideal retention period. It may well be that the current practice is perfectly adequate or in rare cases even excessive.

It is important to take a wider view than simply looking backwards: it is necessary to look forward in time as well. Are there processes that are used to set up details for future events? Do things need to be set up in advance so as to trigger some forthcoming process? Just as it was important to consider future data analysis events, the analyst must also consider the associated processes. The business endures over time so both the past and the future must be considered, together with the present.

Errors and Omissions

Processes will obey the 80:20 Rule: 80 per cent of the processing will be supported by 20 per cent of the processes. The remaining processes are of equal importance. Poor analysis will lead to the following problems:

- *a high level of manual intervention;*
- *abnormal situations being processed as if they were straightforward;*
- *distrustful users;*
- *a high and continuing level of modification requests.*

This final section returns to the problem of error handling and exception processing. It takes a slightly wider perspective and also includes any unusual or infrequent situations.

There is a very useful rule-of-thumb which can be applied here. (In fact, it is so useful it can be applied to many different problems and environments and not just process analysis.) This is the 80:20 rule and it is shown diagrammatically in Figure 13.2. What does this signify? The point of this is that there are a finite number of events or situations with which the business has to cope. Most of these (about 80 per cent) will be predictable, standard events which can be easily defined. Given a complete list of all the processes that the business actions it is probable that 80 per cent of situations are supported by just 20 per cent of the processes. The converse is also true, the remaining 20 per cent of situations will be unusual and probably highly individual. They will require the remaining 80 per cent of processes to support them.

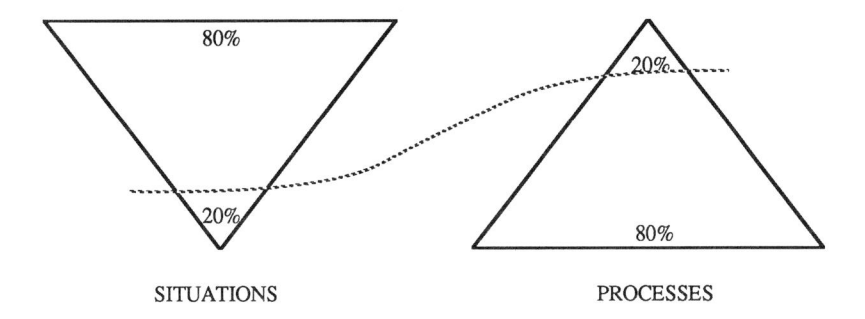

Figure 13.2   The 80:20 Rule. 80% of situations will be handled by 20% of processes and vice versa

General Techniques for Process Analysis

Obviously the first 20 per cent of processes will occur very frequently and will be tightly controlled and organised; as such they will be relatively easy to analyse. The other 80 per cent will occur rarely and may be difficult to uncover purely because of their rarity value. It is this 80 per cent which is in part concerned with error processing. Typically they deal with error correction, omission handling and processing concerning exceptional events. This latter category may occur very infrequently but it is often very important to satisfy these requirements when they arise. The main difficulty with these processes is that they are very easily forgotten both by users and the analyst. There is a tendency to ignore these processes and to assume that as they occur very infrequently then they are not important.

What are the consequences of poor analysis in this area? There is no doubt that it would be possible to ignore these processes altogether and still build a perfectly viable system. Such a system would handle most day-to-day situations and would therefore be of benefit to the users, but it would have major problems. There would be a considerable need for manual intervention, Users of such a system would not be able to rely on the system for anything that was not straightforward. The dangers here are twofold. Firstly, there is always the likelihood of a user failing to notice something unusual and processing the task through the system as if it were perfectly normal. The result of such action could be totally unpredictable and could cause considerable problems later. The second danger is that the users will become very distrustful of such a system and will not rely upon it to do anything right, probably resulting in manual repetition of system functions.

Even if these problems are managed reasonably well there is another danger that may creep up on the designer. It is likely that, as the shortcomings of the system become more and more irksome to the user, there will be a flood of modification requests to improve the usability of the system in order to handle the very process that had been ignored at the design stage. Thus the system will be in a continual state of change with a very high maintenance overhead. The original elegance of the design will rapidly become degraded as more and more patches are applied.

There must be an attempt to uncover these less-frequent processes. It is important to ask what happens when problems arise or things go wrong. We must ask as many 'but what if...' questions as we can think of. This is not quite as open ended as it may seem. Once the users realise that the analyst is interested in the exceptional circumstances as well as the common ones, they themselves are more likely to provide the information needed with less prompting.

Remember the 80:20 rule. It acts rather like an iceberg. The most dangerous part is under the surface!

Chapter 14

Combining Process and Data Models

Introduction to Process Mapping

By cross-checking the processes against the data and vice versa, a number of important points can be resolved:

- *is all the required data modelled?*
- *have all the processes been established?*
- *which relationships are essential?*
- *what are the access points to the data model?*
- *are there ambiguities?*
- *is there any redundancy?*

The position has now been reached where the development and documentation of the data model has been discussed and, somewhat independently, consideration has been given to the processes that the business carries out. There is though, one more set of activities that must be undertaken to be reasonably sure that a complete picture of the business requirements has been obtained.

To follow through with the horse and cart analogy used at the beginning of this book, data analysis has allowed us to define the cart, and process analysis has defined the horse that is to pull the cart. The one thing left to do is to ensure that it is actually possible to hitch the horse to the cart.

The method that is being proposed is effectively self-checking. Establishing the data and process requirements quite separately guards against making omissions or false assumptions much more effectively than would be the case if the definition of the business requirements had been approached from one side only. It should be possible to justify the inclusion of all data by the requirements of some process and, conversely, some process should make use of each piece of data within the model.

Consider the possible outcomes of doing this. Firstly, the results of process analysis can be used to justify the inclusion of each entity type and attribute within the data model. If a fairly academic data analysis exercise is carried out, as discussed, it is quite possible for various bits of data to be included in the model as definite requirements purely due to the fact that the users were freed from operational constraints by the techniques employed. This is not a bad thing in itself, it is far better to have a few spurious data elements included than to miss out pertinent information.

Non-essential data will be a system overhead but is unlikely to be serious. Missing essential data could invalidate the whole exercise. As will be seen later in this chapter, there is a need to establish certain types of process for each entity type. If this cannot be done it suggests that one of two mistakes has been made. Firstly, the data is not required by the business and it should be removed from the data model, or secondly, there is some process that uses that data which has not yet been established. In this latter case, of course, it is necessary to consult further with the users to determine the process and include it. The conclusion of this exercise will establish the reason for holding every entity type and attribute.

Secondly, the correspondence between data and processes must be established from the opposite point of view. Taking each process in turn, it should be possible to find a suitable slot in the data model to hold each piece of data used by that process. Once again, where this cannot be done, there will be two possible reasons for the problem. In the first case, the data model may be incomplete and may need to be extended to allow inclusion of the missing data. In the second instance some additional process may have been invented which is not a requirement of the business. Again, whenever a problem in correspondence is encountered, it is necessary to go back to the users and re-check the findings to discover the underlying cause.

There are some additional checks that can be made at this stage which will be essential for correct system design. It is possible to be quite happy with the data model in terms of entity types and attributes but the data model also describes the *structure* of that data in terms of relationships. Up to now relationships have only been considered from a purely logical point of view.

In pure terms there is absolutely nothing wrong in this approach, but depending on the type of software to be used (an issue which has not been considered yet) it may be very important to distinguish between those relationships that are actually *used* by the business and those that, although logically present, are currently coincidental.

The process mapping discussed in this chapter actually allows the discovery of these highly important relationships and enables the documentation of some detail concerning their usage. This information can be vitally important in later design. Equally usefully, it is possible to discover some relationships that are required to support a Process that has hitherto remained undiscovered. Once again, we are checking our findings in both directions.

Moving one step nearer to design, the process mapping will also yield vital information concerning access to the data structure. The data model has been developed as a somewhat monolithic structure akin to a tall tower. It is important to discover the doors and windows that will allow entry to this structure. These openings can be designed-in at any chosen point but there is always a design trade-off: too many openings and the structure becomes weak, too few and it becomes too rigid and inflexible. By discovering those openings that are needed, the appropriate design decisions can be made.

Finally, there is one very important point. It is almost certain that when the process mapping is carried out we will discover that there are a number of choices about the way we describe it. Firstly, it may be found that there are choices to be made about where to place a particular piece of data in the model. There may be several attributes in different entity types that could serve the purpose. When this occurs the main difficulty is not really to establish which is best – this is usually either obvious or academic – but rather to ensure that there is consistency in the decisions.

It is imperative to avoid holding the same piece of data in one place for one process and in another for the rest. What must be done is to document these choices very carefully and, where appropriate, tighten the definitions in the data model to remove ambiguity.

Similarly, there may be several processes using the same piece of data in different and contradictory ways. Again, there is no magic formula to solve this problem. It can only be done by careful documentation and a consistent approach with adequate review and reassessment of the results.

Thirdly, it is possible that there will be several possible routes through the data model that will support a process. It is necessary to consider carefully which is the best alternative and document that decision so that consistency is ensured in this respect as well.

Finally, it is always possible that whenever these situations arise, there is some redundancy in the data model or the process definitions that should be removed.

The foregoing describes the aim of process mapping, the next section moves onto the actual techniques that can be utilised.

PROCESS NAME	ENTITY REFERENCE						
	1	2	3	......	51	52	53
Input Proposal Details	✓	✓	✓				✓
Calculate Full Premium		✓	✓			✓	✓
Issue Refund		✓			✓	✓	✓
Issue Payment Request		✓			✓		✓
Etc.							

Figure 14.1 A simple Entity Type/Process Cross-Reference Chart

PROCESS NAME	RELATIONSHIP						
	Pol-Cover	Veh-Cover	Pol-Due	Pol-Rcvd	...... Pol-Cust	Cust-Addr	Addr-P'Code
Input Proposal Details	✓	✓		✓	✓	✓	✓
Calculate Full Premium	✓	✓	✓		✓	✓	✓
Issue Refund				✓	✓	✓	
Issue Payment Request				✓	✓	✓	
Issue Cover Note	✓						
Etc.							

Figure 14.2 A Process/Relationship Cross-Reference

Techniques

Each of the techniques proves different aspects of the model:
- *Process/Entity Type cross-reference*
  - *are any processes missing?*
  - *is any data missing?*
- *Process/Relationship cross-reference*
  - *are any processes missing?*
  - *which relationships are used?*
- *Process Maps*
  - *are all entity types used?*
  - *which are used directly?*
  - *which are accessed via relationships?*
- *Attribute Usage*
  - *are all attributes used?*

This section concerns itself with documentation techniques aimed at allowing the notation of the correspondence between entity types and processes with clarity. By this stage the data model has been documented and there will be a list of all the processes and a description of each. Some final documentation is now required that relates these two together.

Process/Entity Type cross-reference

The first documentation tool to be used is a simple cross-reference chart. Figure 14.1 shows an extract from just such a table. As can be seen the entity types are listed across the top of the chart and each process is identified down the side. There is no particular subtlety to this technique: all this does is to ensure that, at this rather superficial level, there is at least one process using each entity type. This is a very quick and easy method and will show straightaway if there are any obvious processes that have been missed. There is a more sophisticated version of this cross-reference chart which will be discussed, but for the moment this rather gross technique will suffice.

Process/Relationship cross-reference

It is also useful to complete a similar cross-reference chart for the relationships and an example of such a chart is shown in Figure 14.2. There is no essential difference between these two charts. In both cases the aim is to ensure that each part of the data model is used by at least one process. If there is no process for an entity type or a relationship then there is either a process missing or those in an entity type or relationship which is surplus to requirements. In either case, there is some further investigation to do.

For the moment the question of attribute usage will not be considered. We will return to it shortly, but first it is important to ensure that each process within the data model is supported.

Process Mapping

A technique of process mapping will be used to describe how each process is supported. Figure 14.3 shows a cut-down process map for the process 'Input Proposal Detail'. As can be seen, the appropriate parts of the data model have been reproduced and the individual steps within the process have been superimposed onto the model. Figure 14.3 gives the process detail in a concise form. To describe the same content in prose is quite long-winded: for example, Figure 14.3 represents the following:

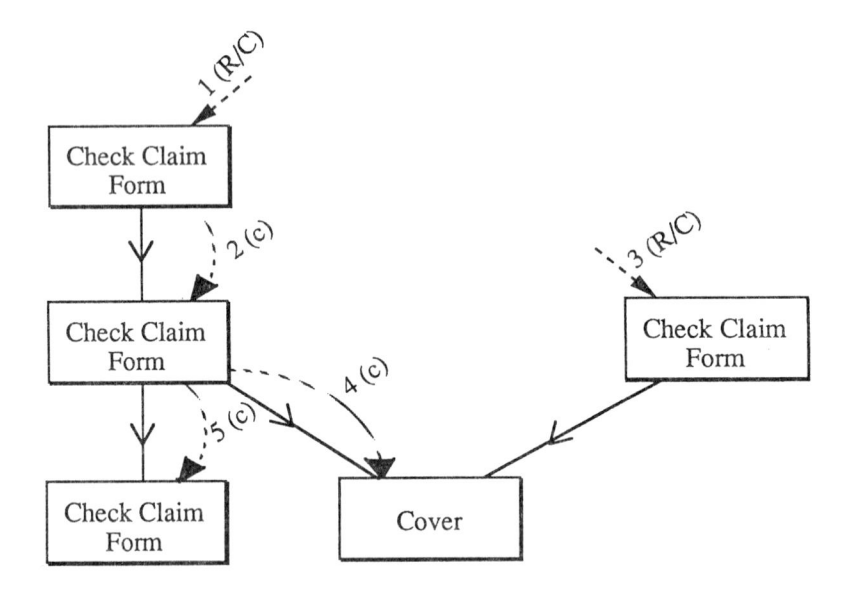

Figure 14.3 A sample process map

Process: Input Proposal Detail.

> Step 1: Read the CUSTOMER entity type to check if the customer already exists. If the CUSTOMER does not currently exist, then create a new CUSTOMER occurrence.

> Step 2: Create a new POLICY occurrence related to the existing or new CUSTOMER.

Step 3: Read the VEHICLE entity type to check if the insured vehicle is currently recorded. If no existing VEHICLE is available, then create a new VEHICLE occurrence.

Step 4: Create a new COVER occurrence linking it to the new POLICY occurrence and the existing or new VEHICLE.

Step 5: Create a new PAYMENT RECEIVED occurrence for the new POLICY.

It takes considerably longer to write the prose version and it also takes much longer to read and understand. The process map technique is an extremely powerful and efficient tool for documenting the detail of the process.

It is also very easy to synthesise a complete picture from a number of process maps. These may simply be superimposed one on another to yield a complete picture of the processing requirements for that part of the data model. This may be done much more easily with diagrams than with text, however well structured the latter may be.

There are two aspects of Process mapping which are worthy of further comment. Firstly, Steps 1 and 3 denote 'access-points' into the data structure and these will be discussed in the next section. Secondly, note the use of the letters 'R' and 'C' on each step of the process. This detail is very important and denotes the type of 'use' that is made of the entity type. This is discussed in the following section.

The basic technique requires the production of a process map for each process that has been identified. It is very important that this exercise is carried out thoroughly and conscientiously and that the results are carefully checked with the users. This detail will be the basis for system design and program construction. Any major errors here will probably be detected very quickly, but more subtle errors may lie hidden for a long time and be difficult to detect.

Attribute Usage

Finally, to return to the question of Process/Attribute correspondence. It is possible to produce a cross-reference chart for processes against attributes, but this can be an extremely long-winded activity and it is difficult with a large data model to maintain concentration! If the basic data analysis techniques in Part 1 have been employed, it will be found that the data will tend to be very fragmented and few entity types will have a large number of attributes. This allows us to take a short cut.

Use a copy of the entity descriptions to carry out the following actions. Take each process in turn, consider each step in the process and tick off, on the entity descriptions, the attributes that are in some way used by that step. Continue with each step and each process. Look out for two warning signs.

Firstly, note each case where only some of the attributes within an entity type are utilised by that step: this suggests that two entity types which should be separated have been combined.

Secondly, once each Process has been considered then skip through the annotated descriptions and look out for attributes that have not been used at all. This serves the same purpose as the earlier cross-reference charts: if there are any unused attributes then either irrelevant data has been included or a process has been missed. Yet another trigger for further work!

Noting Access Points

Data can either be selected directly or via a relationship. Where data is accessed directly then:

* *occurrences must be uniquely identified;*
* *file placement decisions are affected.*

This section considers the implications of access points into the data structure in more detail. Just as a reminder, the standard convention in terms of the process map is shown in Figure 14.4. The convention shows the access as an arrow entering an entity type 'out of the blue'. In Figure 14.4, 'n' represents the step number within the process and the character in brackets signifies the type of access. Why are access points so important?

The significance is concerned with the later activity of systems design. At some point we have to leave the somewhat conceptual activities of data and process analysis and begin to produce an implementable solution. This is, after all, what we get paid for!

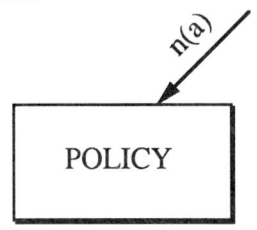

POLICY
n(a)

Figure 14.4   The convention for access points

In broad terms the data can be accessed in one of two ways: either by exploiting a Relationship to find the data, or access can be direct. If a Relationship is used to select data, the total population of occurrences need not be uniquely distinguishable one from another as it is the membership in an *occurrence of the relationship* that is used for selection. This point is more fully explained in Chapter 18. However in order to access occurrences of an entity type directly, there must be some attribute or combination of attributes that allows identification of one occurrence of the entity type from every other. In most cases, if the attributes have been defined carefully this should already have been done, as the Identifying Attributes for each entity type should have been determined. But when designing the implemented record, derived from the entity type, it is necessary to know if it is essential to implement the Identifying Attributes explicitly. It is good practice to always implement Identifying Attributes anyway, but in some cases it may be necessary, for operational reasons, to consider this carefully.

Additionally, it may be necessary to know which records are accessed directly, and which are not, for file placement purposes. For many implementations it is possible to control where records of a particular type are stored within the physical files. This holds true whether conventional flat-files or sophisticated database systems are being used. To help selection within a relationship, it may be helpful to 'cluster' records together so that the number of I/O transfers is minimised. For example, if there is a requirement to access an ACCOUNT related to a CUSTOMER, the only ACCOUNT occurrences that need to be read are those related to the appropriate CUSTOMER. In this instance it would be useful to 'cluster' all the ACCOUNT occurrences related to one CUSTOMER in close proximity to each other, and preferably close to the CUSTOMER occurrence itself. But, if direct access is required, it would be preferable to spread out the ACCOUNT occurrences to minimise contention on the disc and to achieve an even distribution of records.

These decisions may not be straightforward. When all the process maps are considered together there may be contradictory requirements in terms of relationship based access and direct access. Also, there may be several relationships around which the occurrences could be clustered. For the moment, these questions can be ignored. At this point what needs to be done is to note the entity types requiring direct access. The design trade-offs can be made later and, just to reiterate, these decisions are more fully explored in Chapter 18.

CRUD

A CRUD Chart can be developed which indicates how a process uses the data:

- *Create process*
    - *there should preferably only be one;*
    - *it should come logically first.*
- *Read process*
    - *follows the create;*
    - *if no read process then the analysis is suspect.*
- *Update process*
    - *follows Create/Read;*
    - *need not occur.*
- *Delete process*
    - *preferably only one;*
    - *logically last;*
    - *should be no reads or updates following without another create.*

The somewhat cryptic title for this section is in fact not as absurd as it may at first appear. In the second section of this chapter the entity type/process cross-reference charts were introduced and it is now time to delve a little deeper into the usefulness of this technique.

It is important to know *how* each process affects each entity type and there are four basic ways in which this can happen. Firstly, a process may CREATE new occurrences of an entity type. Indeed, every entity type, if it is to be used at all, must have at least one process that causes this creation. Logic dictates that the create process must be the first to operate on a given entity type. There must be a create process even if it comes down to a one-off data load facility. Life is also made easier if there is only *one* create process for the entity type: indeed if there are more than one then these processes should be reexamined to determine if they are the same process under different names. Multiple create processes can cause all manner of control problems later.

Secondly, entity occurrences may be READ by a process. Again this is fairly fundamental: there is little point in storing data if it is never looked at by any process. There must be at least one process that reads the entity type and it must logically occur after the create process or there will be nothing for it to read!

Similarly, at least one read process should normally occur prior to any updating processes. It would seem a little pointless to have a create followed by an update with no use of the initial data content. This is not a golden rule, but generally such a situation requires investigation.

Thirdly, there may be one or more processes that UPDATE entity occurrences. As with the read processes, they must logically occur after the create process, and there should normally be at least one read process prior to the first update although this is not mandatory. There may be a number of entity occurrences, or even whole entity types, that once created are never changed. It is also possible to argue that data should never be updated (that is, overwritten with new values) but rather, a new version of the entity occurrence should be created and held alongside the original. This returns to the earlier arguments about the importance of historical data and time-dependency considerations. If data values are overwritten the audit trail may be lost and it will also be difficult to provide effective historical enquiry facilities. In every case where there is a process that updates data values there should be careful consideration as to whether this is the real requirement or whether a time-dependent view of that entity type would be preferable. In short then, the requirement for an update process is optional.

Fourthly, each entity type may be used by a process that DELETES entity occurrences. In most cases, entity occurrences will not be needed indefinitely. There will come a time when the data should be removed from the system. To facilitate this there must be at least one process that causes deletion of out-dated occurrences. Logically this should be the last process to access an entity type. The create process should already have been noted and the read and update processes should also come before any delete processes. As with create processes it is preferable to have only one delete process for an entity type as this simplifies control over the data and allows deletion of entity occurrences to be carried out in a standard way. Once again, where there are apparently several delete processes, consideration should be given to combining these if possible.

It is very easy to forget about these processes and, although a deletion process is not mandatory, the lack of such a process for an entity type should suggest that this aspect has not been analysed sufficiently and that further work is required.

Hence the title of this section: for each entity type there should be processes that:
    C reate
    R ead
    U pdate
  and D elete
    occurrences of that entity type!

Finally, there must be consideration as to how the results will be documented. It is helpful to use an extension of the entity type/process cross-reference chart (an example is shown in Figure 14.5). By ensuring that the processes are carefully arranged in sequence, it is quite easy to scan each entity type column to determine if the sequence of processes is valid and that all processes have been uncovered. One final point: it is possible for the same process to carry out several different types of access on the same entity type. This will happen, for example, if a number of occurrences are searched in some way and then a new occurrence added to the list. Alternatively, this can happen if an update occurs after several reads. Generally speaking though, if the processes have been carefully decomposed then there will be few, if any, of these situations.

PROCESS NAME	1	2	3	......	51	52	53
Input Proposal Details	C	C	C				C
Calculate Full Premium		R	R			C	R
Issue Refund		R			C	R	R
Issue Payment Request		R			C		R
Customer Addr. Change	R	U	U				
Delete Lapsed	D	D	D		D	D	D

Figure 14.5 A sample CRUD Chart

Combining Process and Data Models   239

Other Process Detail

For each process, the following details should be documented:

- *a short textual description;*
- *frequency, including peaks and troughs;*
- *initiation and responsibility;*
- *start and end events;*
- *preceding/dependent processes.*

This short section discusses the remaining pieces of information to be gathered to complete process analysis. As well as the various cross-reference charts and process maps that have been discussed, a small amount of written detail for each process should also be produced.

First of all, a simple textual description of the process is needed. This is broadly analogous to the entity descriptions and states in business terms and in plain English the nature of the process and its purpose. This need not be anything too fancy, a few well-chosen phrases to allow a user to check easily our understanding is sufficient.

Secondly, some statement of frequency is useful. Just as it was necessary to collect certain statistical detail for the entity descriptions, so it is for the processes. There is a need to know how often the process occurs and whether this is an even distribution or if there are any peaks or troughs which should be noted. At this point it is useful to check if the frequency of the processes affecting an entity type – especially those that create entity occurrences – ties in with the growth estimates made for the entity type. This will yield a degree of confidence in the accuracy of the statistical findings.

Initiation of the process should also be considered. Which part or parts of the organisation have responsibility for ensuring that the process is carried out? It will be important in design to know if all the process occurrences are carried out in one area by a few people, or if the process responsibility is very dilute with many different parts of the organisation involved.

Fourthly, establish and state the events that happen either as start events causing initiation of the process or as end events, the events that are a direct cause of successful completion of the process. This event-related detail assists in placing the process in its correct context and again will assist in the design exercise.

Lastly, restate some of the other processes that have a direct effect on, or are directly affected by, each process. It should be possible to extract this from some of the process dependency charts. But a simple statement of the processes that must precede each process, and those that are wholly or partly dependent upon it will again make user checking more straightforward. Process description, process maps, cross-reference and 'CRUD' charts are also shown in the Appendix.

Checking the Results

Check all the documentation thoroughly and look out for:

- *inconsistencies;*
- *where an entity type is accessed in different ways, try to prioritise these requirements;*
- *are all details complete?*
- *are all defined relationships necessary?*
- *redundancy.*

All the basic techniques needed to master both data analysis and process analysis have now been discussed, and all the different pieces of information to be gathered have also been considered. At this point it is useful to have a small review of what has been covered and how the different threads of the exercise can be pulled together.

What have we got so far? As output from the data analysis there are the data model diagrams and the Entity Descriptions. All this should, by this stage, be fairly complete and firm. A stable and accurate data structure should have been developed and the details within the entity descriptions should be accurate and reliable. The textual definitions of each entity type should be unambiguous and the examples should be of a high quality and hang together throughout the text. All statistical and volumetric detail should be complete and consistent. Hopefully, there will by this stage be good user understanding of, and agreement with, both the data structure and the descriptive text.

There will also be an equal volume of documentation from the process analysis. This will consist of a description of the broad business areas, process breakdowns or dependencies showing how the individual processes fit together to service that business area, process maps to show how the processes use the data, cross-reference charts of several types, CRUD Charts and process descriptions giving a textual description and other detail about each process.

In total this is a quite considerable quantity of documentation. The first thing to do, therefore, is to try to ensure its own consistency. Mistakes can creep into large documents very easily and the credibility of the work may be damaged if any obvious 'goofs' are not removed. So a thorough examination of the documents is necessary.

Does it contain all that it should contain? Does it address the correct objectives? Is it consistent? All these things should be borne in mind by the analyst. This task is demanding, time-consuming and boring – but it has to be done!

To complete this chapter, then, let us consider some of the problems or contradictory details that may need to be resolved. Firstly, the way in which an entity type is accessed may require some resolution. In an earlier section of this chapter there was discussion of direct and relationship based access and this information will, in part, determine the physical arrangement of the entity occurrences. But it is quite possible, when examining all the processes accessing an entity type, that a number of different access paths to the entity type will be found. Where this is the case an attempt should be made to prioritise the different requirements so that, when design trade-offs are made, the requirements are clear. This is normally fairly straightforward if the statistical detail for each process has been gathered. By simple addition it is possible to determine which is the most frequent type of access and, if relationship-based, which relationship is used. Note the findings on the entity descriptions.

There are a few other types of problem that have been mentioned already which are worth re-stating in order to round off the whole picture. Firstly, recheck that all the relevant details for each entity type have been obtained and that there are appropriate processes (using the CRUD charts) to access them.

Secondly, the process maps should show the essential relationships. If there are others then there is a need to do more research to determine if the relationships are necessary. Have all the attributes been used? Is it clear why each attribute has been placed in a particular entity type?

Is anything missing from the documentation? That is, are there enough processes for each entity type? Are there processes that cannot be supported *via* the process maps? Is anything redundant? If processes can be satisfied by different parts of the data structure, there may well be some redundancy that can be taken out. Is it clear where each piece of data is to be held? Each entity type should be exclusive of each other entity type if the analysis has been done correctly.

It is sometimes useful, where possible, to enlist the help of a colleague to do this checking. This guards against false assumptions creeping in. It is very easy to produce something which, although clear and unambiguous to the writer, can be interpreted in different ways by different people. Enlisting the help of a fresh mind can turn up all manner of questions for resolution!

This brings us to the end of another part of the book. We should now have mastered the basic techniques of data and process analysis and have some feel for how to utilise them, but in order to get the most out of these approaches we need to have a plan of attack for any typical project.

Part 4 of this book attempts to do just that!

Part 4

Methodology

Chapter 15

The Method Described

Data Analysis in Context

This part of the book is intended to provide some guidance as to how data analysis may be applied in practice and, more specifically, how it may be combined with other system development methodologies. This is not an attempt to provide a full systems development methodology, rather it concentrates on data analysis as a body of techniques and provides a detailed method for applying the techniques themselves. Many of the current methodologies include at least some element of data analysis. The rules presented in this part of the text should embed fairly easily into such a methodology thus enhancing and filling-out an existing approach rather than changing it.

The systems development process contains a number of fairly universally accepted stages. Figure 15.1 lays out these stages and the major tasks within them. As can be seen data analysis and process analysis are techniques which apply primarily to Stage 2 – Systems analysis. There are a number of techniques available for formalising the latter stages, for example structured programming in Stages 4 and 5. These techniques help to produce a good system but do not help to carry out the analysis in a clear and structured manner. If the area under study is not clearly defined (that is, what the business actually does), a satisfactory end-result will not be produced.

It is true to say that until recently there has never been a clearly defined set of tools which could be used at this very fundamental stage. Data analysis and process analysis together provide these tools. It is important to bear this in mind when attempting to combine the following method with an existing methodology.

There are other techniques which may be used in combination with data analysis, the two most widely-known being Data Flow Diagrams and Entity Life Histories. These techniques are not reviewed in this book. The techniques themselves provide an alternative means of documenting the way in which data is used by processes and do not add (in the author's opinion) to the methods outlined earlier. But different people prefer different methods. Your installation may enforce standards that include them, or your favoured methodology may promote their use. Do adopt these techniques if

STAGE	TASKS
1 Familiarisation	Business Overview Major Functions Definition of the Problem Area Identification of Potential Solutions
2 Systems Analysis	Definition of processing) at the business Definition of data) level Impact of corporate strategy Identification of Major constraints Review of potential solutions
3 Solution Evaluation	Define Scope of solution (which processes and what data) Type of computer system (Batch, On-line etc) Type of Hardware (Mainframe/Micro) Type of Software (Flat-file, Database) Specific Hardware and Software to be used Cost/Benefit analysis
4 Systems Design	Detailed system specification Program specifications File/Database design Screen/Report design Identification of manual procedures
5 Systems Production	Program coding/testing Creation of files/database Operations procedures User procedures
6 Implementation	Data take-on Acceptance Testing User training Parallel Running
7 Post-Implementation	Performance Tuning System support Enhancement/Modification

Figure 15.1   The main stages in systems development

you find them useful. In fact, use any technique which helps *you* to document the business requirements clearly and unambiguously.

This chapter breaks down into two separate sections. The first section describes the basic elements of the method, discussing each part of the method in turn. This is followed by a quick and easy cross-reference between the phases of the method and the basic techniques that have already been discussed. This should be used as a rapid guide not only of what to do, but how to do it, and where to find the detailed exposition.

Before embarking on the detailed phases of the method, it is useful to put some context around the philosophy underlying the use of the method itself. The first thing to bear in mind is that this method is not cut and dried. There are many different ways of organising any project and a number of these will produce good results. That, which is promoted here, is a valid general approach which will allow most people to work in a consistent and disciplined manner.

The following method is by no means the *only* way of achieving a well-run project. It is, however, a proven method which is known to be reliable.

This brings up another point. The method should not be regarded as a set of rules which must be obeyed. Every project in every installation is different and the most any methodology can provide is a set of guidelines with general applicability. Do not misinterpret this. The validity of the guidelines is not reduced; if anything, by approaching any methodology with a questioning mind, the value of using it is increased. A methodology will not guarantee a good result, but it will assist a thorough approach. By following rules blindly it is quite possible to end up with a bad design ..... but at least it will be a *thoroughly* bad design!

Finally, the method is designed to be adapted and revised by each user. If a phase of the method seems in practice to be in the wrong place for your own preference or systems development standards, then change it! Any methodology is merely a synthesis of different people's experiences in different situations and it should continue to evolve. As will be seen, this reviewing is explicit in the method and is highly valuable. We must learn from our past experience or we will not implement the quality of systems that users expect.

Defining the Scope

By setting the scope 'up front' it is more likely that the project will reach its final target. This cannot be too clearly defined and clarification will need to be carried out as the project progresses.

Once the scope has been set, an outline project plan should be developed identifying expected timescales and resources. This can be used to ensure that any training/familiarisation that is required will be carried through.

The first step in establishing any project is to determine exactly what it is that is to be achieved and how to approach the solution. It is necessary to draw boundaries around the project and, although these should not be too black and white, it is important to delineate the area to some extent. Why does this need to be done at the outset? It is true that as any project progresses the target will become clearer and more rigidly defined, but if the target is not defined to start with there is no certainty of heading in the right direction. It is becoming increasingly necessary to control project timescales and budgets with greater precision. By establishing the scope of the project right at the outset it will make project planning and control much easier and will also prevent the project from becoming side-tracked down possibly unproductive blind alleys. At the same time, however, it must always be possible to change these boundaries as understanding of the subject area increases. It is always possible that certain side-line issues may need to be fully incorporated into the project, or, conversely, that they should not be included and that the boundaries need to be redrawn to cut these out completely.

Once the scope has been set with reasonable certainty, it is possible to begin to form an overall strategy, an overall plan of action that will act as a guide towards successful completion. It is not usually possible to make definitive plans at this stage but it is possible to break the task up into its major phases and estimate approximate resources, timescales and other costs for each of these phases. Our masters will always want this level of detail as a minimum and it is not unreasonable for estimates to be supplied at this stage, but beware of making firm commitments at this point. Any estimates made at the outset will be no more than that – they are purely estimates.

With the overall plan it will be known which techniques are to be used, when they are to be used, and who will be required to use them. This means that a start can be made to assess the training or familiarisation requirements that will need to be in place before the project sets forth in earnest. This is very important. Time spent now ensuring that all concerned are aware of the project plan and its techniques will pay dividends as the project progresses. Additionally, it helps to ensure that the personnel who are to actually do the data analysis, process analysis, design and system production are trained and prepared to play their part at the the appropriate time.

Establishing Control

In order to get the project off to a good start it is important to establish the right atmosphere:

- *establish a steering committee of senior managers;*
- *establish the initial project team (quality and quantity);*
- *identify and build a rapport with the main user contacts;*
- *plan the first phase.*

Many data analysis projects seem to gobble up enormous resources and then fail to deliver the goods. Why is this? In most cases this would appear to be due to the way the project is controlled, or more precisely, the way in which it is allowed to be controlled. Data analysis requires a fairly large amount of effort and resource 'up front', before anything really tangible is seen. To be successful, then, it is absolutely essential that the project leader obtains considerable commitment to the exercise and at the highest possible level. This point has been discussed earlier but it cannot be emphasised enough. Having set the scene and established what it is that is to be done, there is now a considerable PR exercise to sell the project and the technique. Considerable user involvement will be needed throughout the life of the project and this can only be guaranteed with the backing of the most senior management.

Getting initial commitment is one thing: keeping that commitment going is quite another. To do this, clear and frequent communication is required and people must be aware of progress as it is made, and even of problems or potential problems, so that appropriate action can be taken. One very good way of doing this is to establish a Steering Committee on a suitably high level. Such a committee, meeting monthly perhaps, would have a senior manager from each part of the organisation as necessary, plus the data analyst, and it should be chaired by a more senior manager than the others. This person should be thoroughly committed to the exercise and should have sufficient authority to commandeer resources if these are not forthcoming voluntarily. Such a committee, properly constituted, will ensure that full and proper communication is achieved between all parts of the organisation and it also helps to formalise responsibilities.

These responsibilities are bi-directional though. Not only is there a responsibility on the users to assist with the project, there is also a clear responsibility on the data analyst to report progress honestly and to inform the committee of any foreseeable difficulties or problems.

Whether there is a steering committee or not, and also regardless of its make-up, it is still important to ensure that senior management are aware of what is happening. Senior management set the work priorities and if the data analysis exercise is perceived as an academic DP activity with little business relevance the project will simply not get anywhere and is doomed to failure. This is a two-edged sword! If you get all the commitment you could possibly ask for and still make a mess of the project, all those senior managers are not going to be very happy! A large data analysis project is not for the faint-hearted!

It is now appropriate to formalise the project team and get the appropriate staff on board. Try to keep the project team as small and tightly-knit as possible. A good flow of communication between team members is essential and too large a team will make this difficult. Breaking the task up into manageable portions can also be problematic. If there are too many staff it is tempting to break the project up into pieces that are too small to allow each person to have a clear perspective of where their work fits in with the overall picture. As a general rule it is preferable to have too few people than too many. The project may take a little longer to complete but it is more likely to yield a consistent and comprehensive result.

This is the time to formalise which actual users are to be the main contacts. It is now right to establish a *rapport* with those individuals and to give them a reasonably thorough introduction to data analysis and how the project is to be run. It is important to give these individuals a clear idea of what is expected from them and how much time they need to devote to the project.

Do be honest at this stage. There is a much greater chance of getting their continued co-operation if they are given honest estimates of their involvement. By under-estimating, accidentally or deliberately, you will just be storing up problems for the future. Take this opportunity, also, to find out what the major problems are for these users. Discover the key business issues and understand them. Although the project should not be overly swayed by the problems of the day, it is important to ensure that they are understood and that they are addressed.

Now plan the first phase of the project. This phase consists of that body of work that leads up to the production of the first data model. This first model should be fairly sophisticated. There is little point in producing a model that is so superficial that it says very little and it may take several man-weeks of effort to establish this first model. This first phase then, involves gathering all the initial information that is needed, producing interview plans or meetings schedules with the users and allowing sufficient time to document the results to a good standard. All the major entity types and all the main processes should be safely documented and understood. You cannot expect to dot every 'i' and cross every 't', but you should end up with documentation that will be fundamentally correct.

To produce the plan, this should be broken down into its constituent activities and time and people resources allocated to each part. This is no different from any other planning activity and any project planning methodology with which you are familiar can be used to assist in the production of the plan.

Once there is a plan, the initial set-up is complete and the project may begin in earnest!

Initial Information

This initial phase should include information from a number of sources:

- *the main user contacts;*
- *senior managers;*
- *publicly-available company information;*
- *existing systems and procedures;*
- *internal/external audit.*

The size of this phase of the project will depend greatly on the background and experience of the people involved. For example, a consultant undertaking his first job for a client will have to begin very much from scratch with little knowledge of the client and, in extreme examples, only slight knowledge of that business sector. On the other hand, a loyal servant of the company with many years' service will know the background inside out and will be able to get down to a detailed level after only cursory checking of the basic area. Neither situation is necessarily the better – the consultant may well bring the benefit of an open mind and be able to cut through layers of company culture that has no real bearing on the problem, and the experienced company man may believe in company attitudes that have become ingrained over time. In both cases, therefore, it is necessary to do some high-level information-gathering to ensure that the right approach is being taken for the right reasons and not just because "it seems obvious".

Once again the basic information-gathering techniques needed here have been covered in earlier chapters. At this point, start at the top. What is it that senior management want to be addressed? Are they committed? Is there a clear understanding of the problem area and where it fits within the business as a whole? If these questions can be answered with clarity then proceed.

It is possible to gain a good initial 'feel' for the company from two sources. Firstly, higher management can provide a broad description of the marketplace, the company and its objectives and problems. Secondly, similar detail can be obtained from publicly available company information such as annual reports, sales brochures, product descriptions and the like.

All these are valuable because they provide an insight into how the company views itself and to some extent its management style. If there is unfamiliarity with the marketplace this will also have to be researched. Who are the main competitors? What products do they offer? How is the marketplace and the various companies' market-share changing? There is no need to become an industry pundit by this exercise, but during the later discussions with management at various levels, these factors will surface and it is important to gain an early perspective on this so that the implications of other information will be understood. At this stage it may be useful to produce a high-level 'quick and dirty' model just to ensure that the scope has been set correctly and that the project is on the right track. This should be used to check out initial ideas and not as a constraining influence on later models.

With this high-level background in place, it is time to begin to interview the main user contacts, moving fairly rapidly across the subject area trying to deepen understanding in stages without charging off down blind alleys to the *nth* degree of detail. These first interviews need to be quite highly structured and this takes not a little skill. It is useful to have a clear list of the main areas that need to be addressed and take each in the most *natural* order possible. This needs emphasis. If, for example, you wish to discuss Trading Currencies, Company Products and Countries, do not insist on this order. If the discussion on currencies naturally leads on to countries then so be it. Let the discussion flow on and return to products at a later stage. Additionally, minute detail is not necessary or desirable at this stage and the discussion needs to be kept moving at all times. Once someone gets onto their favourite bandwagon, tact and diplomacy become important. The discussion needs to be gently steered away from the subject. If it is boring everyone else to tears – come back to it later in a private interview!

There are other important facilities for gathering information and these should be used in tandem with the interviews. Existing computer systems and internal company manuals or procedures can provide a wealth of detail and should be exploited – but do be aware of the problems, discussed in Part 2, that are inherent in the use of such material.

Another useful source (and one that is greatly under-utilised) is the audit function of the company, both internal and, where possible, external. Audit departments typically have much information on the information flows and procedures of the company and such readily-available documentation can save many hours of discussion.

By making intelligent use of all these information sources it is possible to develop quickly a thorough understanding of the company. This in turn will lead to the identification of the main entity types, their attributes and relationships. By the same process it is possible to gather details concerning the processing requirements and also to get an idea of the volumetric and statistical detail required. This should now be structured and written down.

The First Model

Getting the first model right is very important. It should cover the following:

- *approximately 80 per cent to 90 per cent of entity types should be known;*
- *the level of detail will vary but the central concepts should be well-understood;*
- *avoid bland descriptions;*
- *cover the major processes.*

At long last you are in a position to produce your first data model for the project! In many ways producing the first model is the hardest part of the exercise for, by following all the techniques described earlier, the first model will be quite a sophisticated piece of work. Typically, between 80 and 90 per cent of the entity types should have been uncovered and the whole set of documentation should fit together reasonably neatly. It will not be perfect. There will be many detailed areas of doubt and uncertainty, but at least you should now know what it is that you don't know! This is the acid test. If such an entity model can be produced then it will almost certainly be a firm foundation on which to build. If not, then it is back to the drawing-board!

A major point of discussion here is the level of detail required on this first model. The first thing to remember is that, if the subject area is large, the level of detail will not be constant and the model should only reflect what has been learnt: no more and no less. Do not try to add extra detail in the light areas if this is not known, nor should there be an attempt to take a more superficial view of the detailed areas for the sake of mere presentation.

This can be tested. It should be possible to list out the areas or concepts that have in some way been modelled. By drawing a graph of how central or critical each concept is against the perceived level of understanding of that concept, the result should be a fairly straight line as shown in Figure 15.2. In other words, those concepts that are absolutely central to the whole area should be well understood but those less-important fringe areas will be less clearly defined. Try it!

If there are some central concepts where understanding is limited then these areas are of the highest priority for further investigation. Otherwise concentrate on those concepts towards the left-hand end of the graph, the aim being to reach a stage where the graph is tending towards the horizontal.

The descriptions at this stage will not be brilliant. There is a danger here: it is very tempting to give a very bland description especially for those less well-understood entity types. Try to avoid this! Be concise but do attempt to describe fully the current understanding of the entity type and make use of the 'Questions' section for each entity description to point-up those areas where further understanding is required.

Do not forget about processes either. Although the emphasis in this book and in this method is on getting the data structure correct, there is no good reason for duplicating the information-gathering by repeating it for process analysis. The results of the user interviews and all the other information sources can be used to provide process detail and this should be documented every bit as rigorously as the data. The only thing that cannot be attempted at this stage is a thorough cross-referencing of the data to the processes. This should be left until the analysis work has been all but completed. Similarly, ensure that the level of understanding of the data runs ahead of the understanding of the processes. This is fundamental. If the process analysis is progressed too quickly there will be a tendency to start designing the data to fit the process – the very thing that data analysis is trying to avoid in the first place. This is shown graphically in Figure 15.3.

Although at the end of the project the level of understanding of both data and processes should be the same, in the earlier stages it is to be expected that data analysis will take the lead.

So there is an initial model – now it's time to put your head in the lion's mouth!

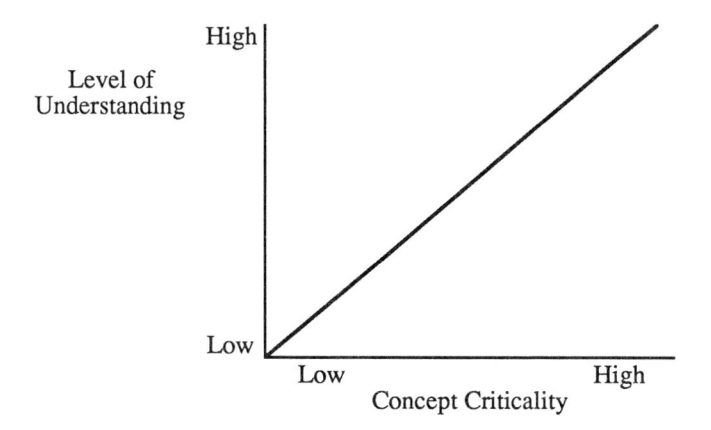

Figure 15.2   Criticality of concepts against level of understanding

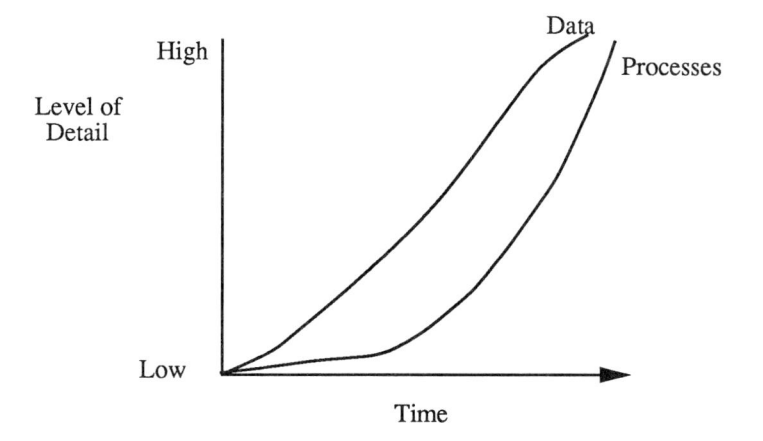

Figure 15.3   The level of understanding of the data should precede that of the underlying processes

Reviewing the Model

> *When reviewing and re-issuing bear in mind the following points:*
>
> - *these sessions will take a long time and will be hard work;*
> - *take thorough notes of all points raised;*
> - *consolidate and incorporate all points raised if necessary taking several passes through the comment and the model;*
> - *carry out a thorough consistency check before re-issue.*

Having produced the first set of documentation this must now be reviewed. User understanding of the documentation is of paramount importance here. Do as much as possible to ensure that the users do understand the implications of Relationships, Entity Types and Attributes and that they are also quite happy with the process modelling techniques used. It is much easier, of course, to sweep the users along with you if they do not really understand what you are saying. However, the more fully they understand the more likely they are to offer constructive comment and criticism. These review sessions, if run correctly, will not be easy. If fact, the harder the better! It is, therefore, important to allow sufficient time for these review sessions. It is unlikely, in a first review, for a group of users to discuss in excess of five to seven entity types in an hour. If the data model has 100 entity types in it, that adds up to a considerable number of man-hours.

The data analyst must pay attention to all the comments and criticisms made. The contributions must be noted for later consideration. There is usually insufficient time within the review sessions to follow up all the implications of each comment, so good note-taking is essential even where the idea may not seem particularly relevant. It is often the case that a comment made at one point in the model will have implications either earlier or later, and these will need to be carefully assessed.

Following these sessions it is time to start consolidating and rationalising the comments gathered. It is necessary to take each comment in turn and assess its effects on the documentation and, where necessary, to assess the joint effects of a number of comments that may have been made at different times by different people in different contexts.

It is often necessary to take up some of the points at a later date with the relevant users if doubt still remains as to the implications. The easiest way to coordinate this work is to use a working copy of the documentation and actually write onto it the changes caused by each point. This allows a more consistent record of the changes to be maintained and minimises the difficulties of assessing the effects of combinations of changes. In this way, it should be possible to work through the list of comments, ticking off each as it is incorporated into the main documentation. It can sometimes take several passes through the comments before each is fully incorporated but there is nothing wrong in this.

Once all the changes have been carefully considered and incorporated into the working copy of the documentation it is almost time to re-issue the model. All that needs to be done before this, is to make a final pass through the working copy to ensure that the different parts of the model are consistent and that it all still hangs together. The simplicity of this task is determined by how well the changes were incorporated in the first place. If this was not done particularly thoroughly, the consistency checking will be more difficult and the risk of a poorly-presented document is increased.

Having done all this the final re-issuing of the document becomes fairly mechanical especially where wordprocessors or similar automated means are used to edit the original. However this process itself can still take considerable elapsed time for which suitable allowance should be made. As a general guideline it is not unusual for a data model of about 100 entity types to need about three months' elapsed time between initial production and re-issue.

Iteration

> *Iterative design needs clear control to prevent it from becoming an unending process. The aspects requiring most control are:*
>
> - *How many iterations? Generally speaking five or six iterations should be allowed for. It is also possible to judge the amount of work needed by the level of change to each issue;*
> - *How often? The outline plan should allow a longer time for the first re-issue. This will gradually reduce as later issues require less fundamental change;*
> - *Control: clear objectives should be set at the outset which will allow progress to be judged and also enable the exercise to be brought to a conclusion. These objectives need to be communicated to, and agreed with, all parties involved.*

Is it possible actually to complete a data model? Probably not. By using an iterative design method to increase understanding gradually it always seems to be the case that, at any point, there are always further questions that can be asked and further areas into which the model can be extended. It is quite possible to get trapped into this method of working – 'Paralysis by analysis', as it has been dubbed – but the cycle of 'produce a draft, criticise, consider and redraft' is useful. The questions that remain then, are concerned with: "how many times to go round the cycle?" "how often?" and "how is the process controlled?"

The 'how many?' question is governed by the law of diminishing returns and is shown in Figure 15.4. The basic point here is that the more iterations there are, the less understanding actually increases. A good data analyst with a sound understanding of the business area will rapidly develop an accurate and complete model. After about five or six iterations changes tend to occur only at the most detailed level and the usefulness of the model for subsequent design is not enhanced. Typically, three to five iterations are needed to bring the model to a good stable state and, although there will still be questions that could be answered, the basic structure will be stable and may be used as a firm basis for further design. Not that the model should be left to vegetate! The model should continue to evolve to meet changes in the business and this is an unending on-going task. The skill comes in recognising when the model is sufficiently stable and in allowing adequate iterations at the planning stage. Two clues can be used to answer the recognition problem. Firstly, what is the general level of change? If this is merely improving the quality of the statistics and honing the entity descriptions without fundamental change there is probably little point in delaying detailed design work.

Secondly, it is often noticeable that a certain amount of 're-invention of the wheel' starts to creep in after a while. It will be found that some detail is being changed in one issue and then changed back in the next. A certain amount of fuzziness will always have to be tolerated and this re-invention is symptomatic of areas that are not going to become any clearer regardless of the amount of analysis carried out. Don't worry about it!

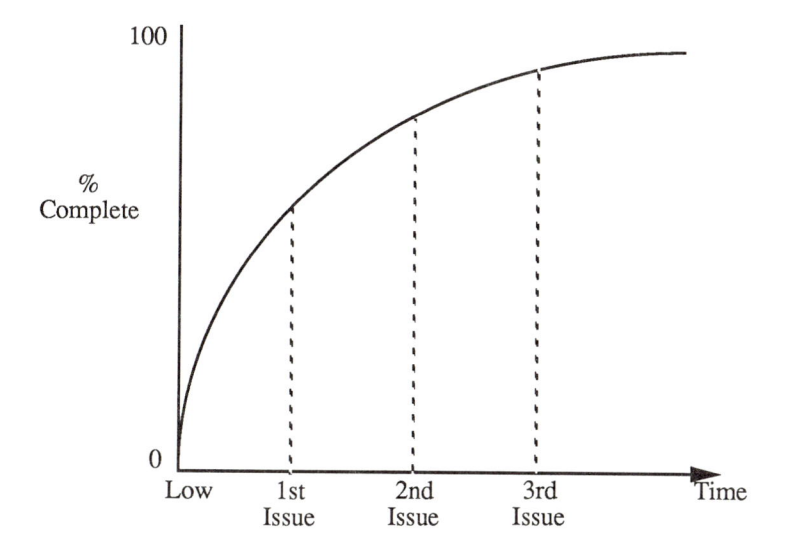

Figure 15.4 Completeness of model over time

There is also a need to set some guidelines on 'how often' the documentation is to be re-issued. This will depend on how many people are involved and the availability of user representation. Generally speaking it will take an average of six weeks per iteration but this is not an even distribution. The first draft of the model and its review will take perhaps three months to complete but as the work progresses each cycle will get quicker and quicker. Once a firm basis has been established it will be possible to fill in the details much more quickly and thus speed up each iteration. These can only be general guidelines: each organisation and each project is different and it is not easy to assess accurately the time-scales involved until the first iteration is complete.

The most straightforward way of controlling this is to set clear objectives for the work that is to be done. It is all too easy to allow a data analysis exercise to go on and on indefinitely. Academically pleasing as this may be, it is necessary to plan and control the work in hand. This can only be done if there is a clear idea of what the end-product is to be used for. These objectives need to be set by the project manager or equivalent and communicated to the project staff. What level of detail is required? How much time is available? How many iterations are envisaged? These questions need to be carefully considered and the implications assessed.

The last point to consider is coordination of the work of the different people involved. If the project covers a large area it is often beneficial to split it into several sub-projects, each being addressed separately. Although this makes the project easier to handle it does bring its own dangers. It is absolutely imperative that there is a clear coordinating role allocated to bring together the work of different data analysts. There are bound to be areas of commonality between the different sub-models and somebody has to take account of these to ensure that contradictory information or redundancy is eliminated. This kind of project can be tackled in the same way though, and plans can be made for a number of iterations of each sub-model, proceeding in parallel if resources allow. Then time can be allowed for a consolidation phase during which period the various sub-models are brought together and their similarities recognised.

Reviews and Talk-throughs

The project team review is intended to achieve:

- *improved technical quality;*
- *highlighting of any inconsistencies/overlap;*
- *the input of a certain amount of fresh thought.*

It should not:

- *be a 'hatchet job';*
- *make the author defensive;*
- *be undertaken lightly.*

So far this chapter has concentrated on getting the most out of the users, but many such exercises will be conducted within a project environment and a certain amount of reviewing and assessment must occur within the project team itself. This section addresses these topics.

First of all, what is reviewing meant to achieve? Mainly, it is to aid communication within the project team and gain input from different individuals. The most useful way of structuring this is to build it in as a formal task at the end of each iteration, immediately prior to re-issuing the documentation. The user-review of the last issue should have been completed and the documentation should have been re-drafted. At this point the document should be subjected to a review with the other members of the project team. This will bring in a certain amount of fresh thought and may resolve some of the outstanding problems or pose new ones. Again, do not be overly defensive of the work that has been done. By involving the project team members it is possible to examine the technical quality of the work rather than its accuracy and this will help to ensure that what is issued to the users is of good quality.

Our colleagues should be looking to satisfy themselves that the document hangs together well, that it makes sense and that there are no obvious contradictions or inconsistencies within it. Additionally, if the project has been split down into a number of areas, it will allow the overlap between these areas to be examined and parallels and similarities to be recognised. These can be very valuable sessions; any problems can be resolved within the confines of the project team, which is less embarrassing than having to admit to a problem in front of the users!

These sessions should be arranged to be as constructive as possible. A 'hatchet-job' is not required. What is wanted is considered and thoughtful comment that can be usefully incorporated.

Get it Agreed

Until the documentation has been signed off, further design work should not be undertaken. It is important not to attempt sign-off too early, though, as the users will feel bullied into acceptance and later changes will arise.

Once the documents have been re-issued enough times and the data model is sufficiently complete, it is necessary to obtain agreement that the work is satisfactory. This should be done as a means of formalising the agreement of the users that the work is accurate and complete.

This is not a time-consuming process, but it is important to stress that it should be done before undertaking design work. The users must agree that they have been understood correctly and that the documentation is as complete as possible. The data model is really a document from the users to the system designers. We, as data analysts, have merely acted as translators along the way. It is this specification which the designer has to turn into a working system. If this cannot be agreed then there is little point in progressing. At the end of the day, the user has to take the responsibility for specifying what they want. We have now reached that point.

On the opposite side, there is little point in trying to get user agreement too early. It should be clear that the work is virtually complete. If not, then another iteration is required before signing-off, and project timescales would have to be adjusted accordingly. Don't forget that the model will need constant maintenance even after sign-off. The business will continue to evolve and the data model will have to evolve with it.

If sign-off can be achieved, the design of a solution can proceed. All that is needed now is one final review of what has been done.

Overall Review

The method proposed here is not inviolate. It should be changed in the light of experience. Questions to ask are:

- *was the project a success overall?*
- *how well did each stage progress?*
- *were resources sufficient?*
- *were timescales met?*
- *were objectives achieved?*
- *improvements?*

The final stage in the exercise is concerned with a quick reassessment of the project to date. By making this explicit it is easier to learn from any mistakes made and to make improvements to the general method for future use. This should assess how well each stage of the project was executed. Were enough resources allowed? Were timescales met, and if not, why not? Were the objectives achieved? What improvements can be made?

All these questions need to be asked and answered. A group discussion involving all the project members is the best way to organise this. Useful feed-back can also be obtained from the users who have been involved. What were their impressions? The answers will allow tuning of the method to suit the particular organisation and its method of working. If part of the project went wrong or was not particularly useful, should this be dropped or changed in some way? Did the method suggested here fit in with any existing system methodology? If not, how can this be improved in future?

Do not be afraid of changing things around. There are many ways of getting to the 'right' answer and it is not the purpose of this book to dictate a particular approach. The old adage 'once bitten, twice shy' should be applied. If some part of the project felt wrong then change it!

Method Summary and Text Cross-Reference

This section provides, in Figure 15.5, a simple, easy-to-use cross-reference between a step-by-step presentation of the method and those sections of text which describe the techniques necessary for that step. This chart can be used as an easy planning aid to define the steps and their sequence. Additionally, it may be used as a handy reference to direct the reader to the appropriate sections of text when guidance on how to proceed from a particular point is required.

In the main, the cross-referencing is to whole chapters. The exception to this is this chapter, Chapter 15. As the sections of this chapter are concerned with specific steps in the methodology, reference to a particular section or sections is made.

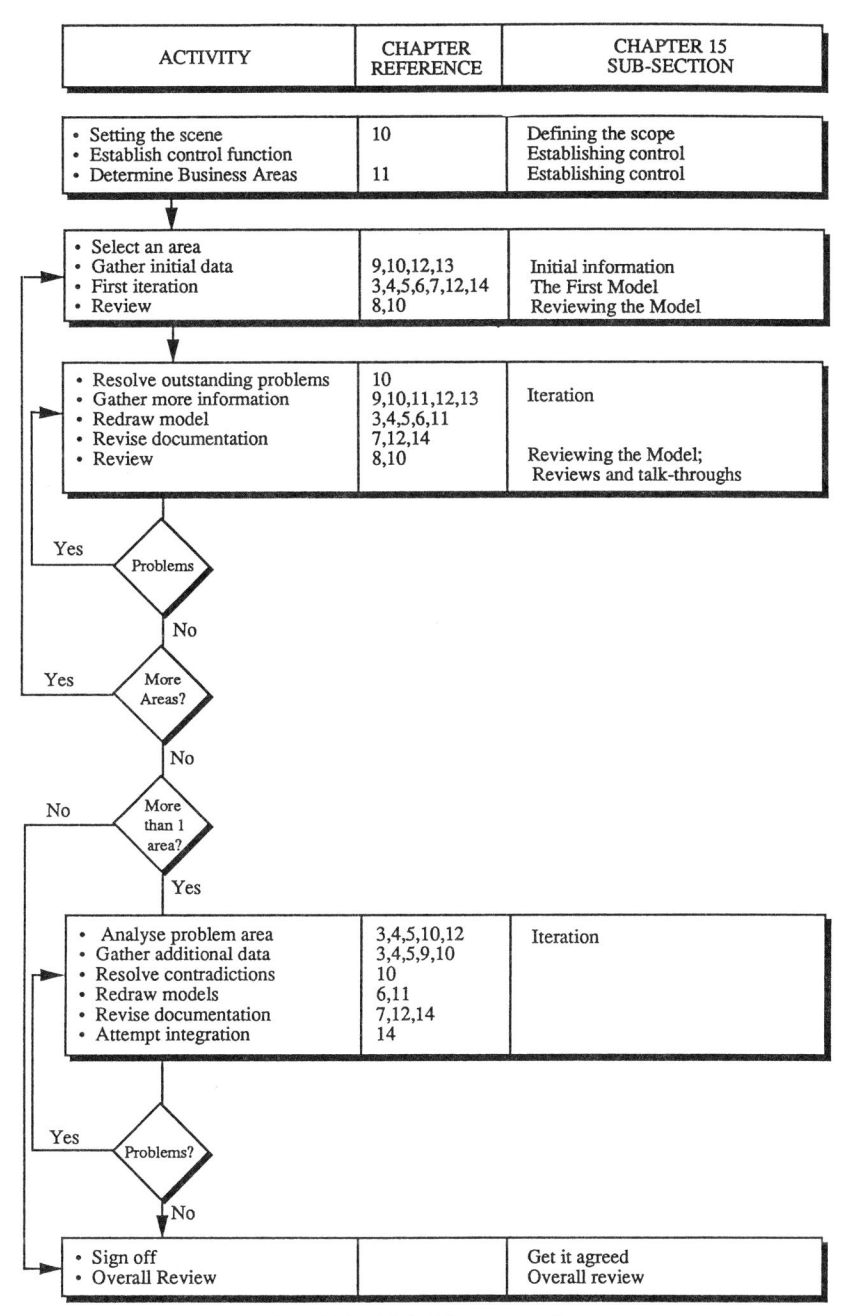

ACTIVITY	CHAPTER REFERENCE	CHAPTER 15 SUB-SECTION
• Setting the scene • Establish control function • Determine Business Areas	10 11	Defining the scope Establishing control Establishing control
• Select an area • Gather initial data • First iteration • Review	 9,10,12,13 3,4,5,6,7,12,14 8,10	 Initial information The First Model Reviewing the Model
• Resolve outstanding problems • Gather more information • Redraw model • Revise documentation • Review	10 9,10,11,12,13 3,4,5,6,11 7,12,14 8,10	 Iteration Reviewing the Model; Reviews and talk-throughs

Yes — Problems — No
Yes — More Areas? — No
No — More than 1 area? — Yes

• Analyse problem area • Gather additional data • Resolve contradictions • Redraw models • Revise documentation • Attempt integration	3,4,5,10,12 3,4,5,9,10 10 6,11 7,12,14 14	Iteration

Yes — Problems? — No

• Sign off • Overall Review		Get it agreed Overall review

Figure 15.5 Methodology cross-reference

Part 5

Using The Model

Chapter 16

Changes to the Existing Business

Introduction

So now we come to the final part of this book. So far it has introduced the basic concepts, described the techniques required, and discussed how a data analysis exercise might be organised and controlled.

This is all very well, but just what use is the data model once it has been developed? The *raison d'être* is a wish to develop better implemented systems and to ease the so-called applications backlog in terms of system enhancement and maintenance.

The data model can be used to address some of these issues. An exposition of the practical uses of the model is contained in this final part of the book.

Of the six remaining chapters, this chapter looks at the maintenance of the model and its use in assessing the effects of change within the boundaries of the existing business area. The world we live in is constantly changing and there will be situations where the way in which the business operates will alter, either because of some external factor or due to some self-imposed reorganisation, or both.

The next chapter is also concerned with change but on a different level. It discusses extensions to the model. These are usually due to a broadening of scope or market place or it may also be due to a fundamental reorganisation of that market to which the organisation must adapt or die. Again the data model can assist in assessing and controlling the impact of such changes.

The three following chapters broach a mammoth topic which really warrants a book in its own right. This is the broader issue of data management and Database Design. Many people, quite rightly, use data analysis as a pre-requisite of database design and it is necessary to give some detail as to how the data model can be translated into a working database implementation.

In conclusion, the final chapter pulls together some of the underlying themes that have been presented and suggests some future directions for the use of these techniques and systems development in general.

Returning to this chapter, how can the data model be used to our benefit when the business changes?

How to Use the Model

The data model provides a bridge between the implemented system and the business. New requirements can be tested against the model to establish the degree of change for both processes and data. This must consider:

- *changes to data (entity types and attributes);*
- *changed processes;*
- *the effects of the changes on existing processes and data.*

No business ever remains static and it would appear that, as life becomes more complex, the speed of this change gets quicker and quicker. Designers of computer systems need to be able to respond to these changes with equal speed and confidence. The development of a data model can assist with this.

This chapter considers changes within the existing scope of the business but even in this context there are two types of change that should be considered. Firstly, there are extensions to existing services. For example, in a banking application, a new type of loan service may be instigated. The bank will already be in the business of lending money and will doubtless have systems capable of processing these transactions. The question here is whether this new type of loan can be handled in the same way or whether some additional work is required to support the new development, and if so, how much.

The second typical area of change is where existing processes are re-organised for some internal or external reason. This may be due to the company centralising or decentralising some operations or it may be imposed externally due to a new legal monitoring process or other restriction.

Although it is useful to distinguish between these two types of change, for current purposes they can be treated similarly. The starting point is that something has, or is about, to change and it is necessary to quantify the amount of effort needed to change the current systems to handle this change.

Many of the causes of poor system maintenance are due to the widespread failure of computer installations to maintain, or even produce, a clear unambiguous description of the system at the logical business level. This logical specification is the bridge that shows how the business functions are serviced by the systems. Notification of change always originates at the business level and is described initially in business terms. Without this essential bridge, the maintenance staff have little option but to take 'a leap in the dark' to translate this into system modifications. This results in a tendency to rush in at the systems level and start specifying the change, sometimes with only the most superficial understanding of the effects of that change.

The data model fills this gap and provides this essential bridge between business and systems. The model is a business specification and as such helps to establish the current situation. The new requirements can be applied against the model to uncover the degree of *logical* change that is required. Is there a suitable entity type to hold each piece of new information that may be required? Are the relationships between entity types adequate for the revised needs? In this way it is possible to start to estimate the amount of new data that is required, any additional entity types or attributes that may be required, the source and other volumetric detail for this data.

The same principle can be applied to any changed processing requirements as well. Are they fully supported by the data model? If not then the degree of change required to develop this support can be determined. Once it has been established which processing requirements can be satisfied, then the frequency and number of these processes can be assessed.

There are other effects as well which must be covered. Having determined the changes that need to be made to support the new and/or revised requirements there must be a check to assess what effect these proposed changes will have on the existing processes. Do the existing processes need amendment because of additional data, either entity types or attributes? It is very easy to make sufficient change to allow the new requirements to work and then find that many of the existing processes fail because sufficient attention had not been paid to this important area. This backward look is essential.

Because the logical level changes that need to be made can be clearly assessed by using the data model it is easier to estimate the time and cost required to effect these changes at the system level. This is very useful budgetary and project control information. The changes can be broken down into individual tasks much more easily and the change resourced accordingly. It should, therefore, give the users a much clearer idea of how long the changes will take and will also allow better control of their implementation.

However, this is not the end of the story and there are some essential do's and don'ts which are discussed in the next section.

Don't Short Cut

Do not bend the world to fit the model. Beware:

- *widening the scope of existing entity types (names and categorisation);*
- *include additional examples to highlight revised scope;*
- *revise volumes;*
- *check the process definition;*
- *check for consistency and clarity;*
- *summarise changes made.*

Using a data model to assess the effects of change is not a magic solution. It is no substitute for *understanding* what the users want and analysing their requirements, but it does provide a facility to reflect these changes at a logical level prior to system modification. As with the initial data model production it is important to be as logically accurate as possible at this stage. There is a great temptation to 'patch' the logical model in the same way as we would patch a computer system. This must be avoided. These short-cuts must be resisted and attempts made to reflect the required change as a full and complete logical description.

This may cause the creation of more changes than were at first apparent but to be true to the philosophy this is the only course of action to be taken. For example, there may be an entity type MOTOR POLICY and it could be that the Insurance Company wants to start dealing in Household Insurance. It may well be that the MOTOR POLICY entity type will be able to accommodate the Household Policy details. But to do this the name should be changed to something more general (such as POLICY) and also add a POLICY TYPE entity type to distinguish between these different policies. The short-cut way of doing this would be to leave MOTOR POLICY unchanged and 'remember' that it now had a wider meaning and to embed the policy type within the MOTOR POLICY entity type without defining it explicitly. The former solution is clearer and more correct. In short, do not bend the world to fit the model, bend the model instead!

If the data model is going to have any long-term usefulness at all, it is important that it is kept up-to-date and that all such changes are reflected in the documentation. Firstly, it is necessary to re-draw the Data Structure Diagram to reflect any changed requirements. Secondly, the Entity Type Descriptions need to be revised and updated as necessary.

If the scope of existing entity types is being widened, it is useful to include additional examples to underline and define the greater scope. Also, particular attention should be paid to the volumetric information as this will certainly require amendment and this is very important for correct physical file design.

In addition to the data descriptions, the process definitions and their attendant charts, maps and cross-references will require revision or extension. If the analysis has been done thoroughly there is nothing clever in all this. It becomes a hard slog through the documentation to ensure that all the necessary changes have been included. Few people enjoy documentation maintenance but it is an unavoidable necessity.

Having included all the basic changes there is a need to make a final run through all the documentation to make sure that everything still hangs together and that the initial consistency and completeness of the model has not been lost or diminished. This can be treated in the same way as a re-issue of a developing data model and it requires ploughing through the same list of tasks, consulting with and getting agreement from, the users as necessary.

One final point: it can sometimes be difficult to detect all the changes made to a data model, particularly on the Data Structure Diagram, when it is re-issued. In order to help others to detect the changes, it is useful to issue a summary of the changes that have been made. This list can preface the re-issue and will highlight those points of interest.

So much for changes within the existing scope of the data model. What about changes of a more wide-reaching nature? These will be examined in the following chapters.

Chapter 17

New Business Areas

Extending the Model

There are essentially three steps in extending the model into new areas. These are:

* *develop a separate model for the new area in isolation;*
* *extend the old model to incorporate areas of similarity;*
* *add the remaining new requirements to the extended model.*

Sooner or later a situation will arise where the data model will need to be extended. This may be because a further area of the business is to be included in the study, or, alternatively, the organisation may have moved into a new area of operations thus requiring new systems to be developed.

The reason for the change is fairly irrelevant. What is important is the fact that further work needs to be done. As before, some basic analysis of this new area cannot be avoided. In effect, a mini data analysis exercise should be carried out to analyse and document the new requirements. This is no different from the work described earlier but it is useful to treat this new exercise as a separate, unrelated activity. By jumping straight in and trying to extend the existing model, it is easy to fall into the trap of bending the world to fit the existing model. This must be avoided as it will, at best, be a compromise; at worst, it could be very much in error. By studying the new area independently with no preconceptions there is a much better chance of creating an accurate model of the new requirements.

So, in essence, it is necessary to carry out the data analysis in the way already discussed. There will be a need to consult with the relevant users and to draw up a data structure and entity descriptions to describe the subject area. At the same time it is necessary to pay equal attention to the processing requirements. These need to be analysed and documented in precisely the same way, independent of the existing process documentation.

Finally of course, this mini-model needs to be reviewed and agreed with the relevant users. Again this is no different from before, there must be full agreement from users that they have been fully understood and that this has been reflected in the new model.

This results in two sets of documentation: the existing model and process documentation, and a similar set for the new area. There are now two major exercises to undertake each of which is described in each of the following sections. First, a review of the existing model to allow inclusion of any similar parts of the new area and, secondly, extension of the existing model to add in what remains to form one composite integrated set of documentation.

Revise Existing Structure

Identify those components of the new model that appear to combine with existing features of the old model. Specific issues are:

- *Entity Types/Attributes: add in any new attributes and check the viability of the enlarged entity type and attribute examples;*
- *Relationships: having added in any new relationships between existing entity types, the structure should be thoroughly reviewed to ensure that the structure is valid and that all relationships are valid and are supported;*
- *Processes: there should be a particular check that all processes are still needed or have not been altered by the new entity types and relationships.*

It is important to remember that there are now two independent but complete sets of documentation. All the entity types, attributes, relationships and processes to support the existing business areas are in the original model, and all the entity types, etc, required to support the new area are in the new mini-model.

If the new mini-model is to be complete in itself it is likely that it will contain entity types, attributes, relationships and processes that are already defined in the main model. This is likely because it is highly *unlikely* that the new area is totally independent. It may, for example, be a new product being sold to existing customers. This overlap has been deliberate by developing the mini-model in isolation. This is beneficial: the description of the new area has not been coloured by worrying about what already exists.

However, it is obviously necessary to pull the two together and to determine the degree of this overlap. How can this best be tackled?

It is a fair bet that the mini-model is smaller than the main model so it is less onerous to try to slot the new requirements into the main model rather than the other way around. This should start initially with the entity types and the attributes. For each entity type in the new model, is there one (or more) existing entity type that serves the same purpose? In some cases this will be obvious, in others it will be more difficult, but care is needed even where the correspondence looks straightforward. For example, there might be a CUSTOMER entity type in both models and it would seem reasonable to combine these.

Are they really the same? This can only be determined by looking at the attributes, and perhaps the attribute values. Are the attributes sufficiently similar? This can be tested by combining the attributes to form a combined entity type and judging how well the two fit together. This is always subjective. The definitions of the entity types may help here. If it is fairly certain that both models are referring to the same thing, albeit possibly different facets of the same thing, then this combination can be 'pencilled in' adding any new attributes for the entity type that have been uncovered. The situation becomes more complex where a single entity type in the new model can be incorporated by splitting it across two or more entity types in the existing model.

This needs special care. It is best described by means of an example. Figure 17.1. shows an extract of an Insurance application capable of handling many different types of policy. The POLICY entity type would contain only those attributes that apply to all policies and POLICY TYPE will differentiate between these types of POLICY. POLICY DETAIL TYPE lists each of those attributes that are applicable to only some of the different POLICY TYPES, the extent of applicability being specified in POLICY TYPE NEEDS DETAIL. The final entity type POLICY DETAIL VALUE holds the specific value for a given POLICY for one of these variable pieces of information. This type of structure is a very powerful and flexible way of handling this kind of situation.

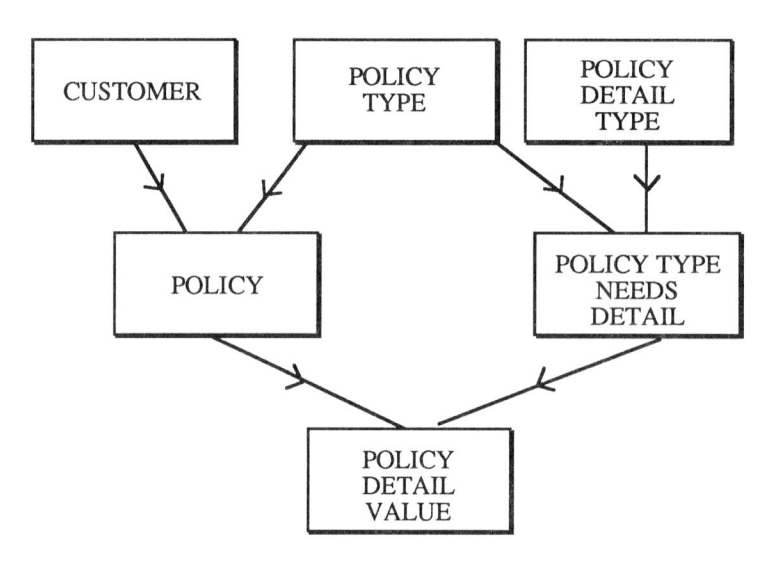

Figure 17.1   An extract from a general purpose model

Figure 17.2 shows an extract from a new model drawn up to reflect the desire of the Insurance Company to move into Travel Insurance as a new product. This model is much simpler having a single entity type TRAVEL POLICY to hold the policy detail. The question is how to combine these two structures. The two CUSTOMER entity types, it is assumed, will combine without difficulty. Furthermore, it is fairly obvious that TRAVEL POLICY should be combined with POLICY in the existing structure, but this has other effects.

```
┌─────────────┐
│             │
│  CUSTOMER   │
│             │
└──────┬──────┘
       │
       ∨
       │
┌──────┴──────┐
│             │
│   TRAVEL    │
│   POLICY    │
│             │
└─────────────┘
```

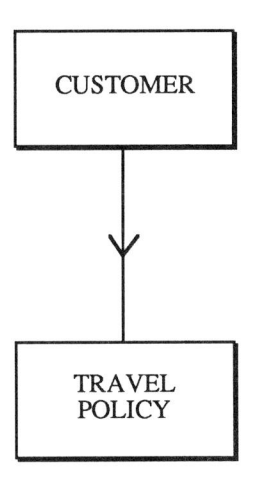

Figure 17.2   An extract for the new 'Travel Insurance' Application

To achieve total integration it has to be established if the POLICY TYPE entity type can handle an additional occurrence for travel policies. The attributes of TRAVEL POLICY need examination and new occurrences of POLICY DETAIL TYPE set up for each attribute that is not used by any other POLICY TYPE. Lastly, is it possible to populate meaningfully the cross-reference entity type POLICY TYPE NEEDS DETAIL for the new 'travel' POLICY TYPE? If all these tasks can be achieved then Figure 17.2 can be incorporated into Figure 17.1 without any change to the existing data structure, but as can be seen, the level of checking required is quite detailed and specific.

By such means it is possible to deal with entity types and attributes and we can now turn our attention to relationships. Equivalent care should be taken with these. Basically, each relationship linking an entity type that has been incorporated into the existing model has to be added to the model and its effect assessed. In many cases, it will be found that new relationships are being added to existing entity types, in other cases there will be an existing relationship that can be used. Care is needed in assessing the effects of these new relationships. It could be that the new relationship causes a reassessment of the existing structure, either by highlighting the need for additional entity types or by prompting a re-evaluation of the usefulness of existing relationships. The only sure way of doing this is to draw the additional relationships onto a copy of the existing model and then consider if this is how the combined model would have been drawn if this had been the starting point.

Remember too, that having added relationships there will be a need to check that there are suitable supporting attributes in the related entity types. If relationships have been removed there may be supporting attributes which can be removed, although the effects of removing an attribute should be very carefully assessed before this is done.

So much for the data structure. Similar checks should be made against the process documentation. Are there any processes that are in both the existing and new models? If so, can they be combined without difficulty? There should be sufficient information about the process requirements to make this decision. The first step is to take each process required by the new area and try to find an existing process that either already achieves that function, or one that is sufficiently close that a minor extension will render it suitable. This should also be turned round to consider whether there are any existing process requirements which are no longer required because the new area has been incorporated into the existing model. This may sound a little unlikely but it is quite possible. There could, for example, be an existing process which produces a report or transfer of data to the new area not previously included in the model. Once the study has been widened to include the new area it could be that the integration of the data structures has removed the need for this transfer as an explicit process. Alternatively the process may have been modified. These changes must be identified and it should not be assumed that existing process requirements will remain unchanged.

Finally, once the revised process definitions have been checked, take a final look back at the data structure. If there have been significant changes to the data structure then there should be a check that the existing process requirements can be satisfied and supported by the revised structure. Hopefully this will present no problems and will simply give more confidence in the design. If there are difficulties then there is a need to go back to the original documentation to see if the source of the problem can be determined and appropriate changes made. If this area is ignored there is a risk of implementing the revised and enhanced structure only to find that large chunks of the existing system stop working!

So much for the effects on the existing structure. In many cases though, only some aspects of the new areas will fall neatly into the existing concepts. Having done all this there may be entity types and relationships remaining which have not been accommodated. The method for dealing with these is described in the next section.

Integration

Certain steps should be taken to achieve the integration:

- *Entity Types: add in the new entity types;*
- *Relationships: add in the new relationships and carefully check the effects of relationships between existing and new entity types; check also for any resulting redundancy and ensure that all relationships are supported at the attribute level;*
- *Processes: add in the new processes, but check that these do not change any existing process descriptions;*
- *Review: check all documentation for consistency and integrity;*
- *Summarise: produce a list of all changes and chart out an achievable path for effecting these changes.*

The remaining entity types, relationships and processes should represent those concepts, and only those concepts, that are totally new to the existing model. These new concepts need to be built into the existing model in such a way that the result is one integrated data model. Unless absolutely necessary, this should avoid just tacking a new chunk of documentation on the back of the existing work. It needs some thought to find the best place to put those new concepts. The question that should be asked is "what is the best way to produce one set of integrated documentation?" and not" what is the quickest way of producing this documentation?".

Bearing this in mind, the initial part of this exercise is quite easy. Any remaining entity types will need to be drawn in and any new relationships added. These relationships may link two new entity types or one new entity type and one existing one. Either way, they should simply be added into the structure. Leaving processes aside for the moment, first concentrate on tidying up the data structure itself. The new relationships that have been added and, in particular, those relationships that link new entity types to existing ones, may have fundamentally altered the structure. They may have introduced duplicity or redundancy within the structure and there should also be checks to ensure that there are the right Supporting Attributes in any existing entity types that have acquired additional relationships. The relationships will need to be rationalised to remove any redundancy and such rationalisation should be carried out where necessary.

Having got the data structure itself sorted out, attention can be given to the processes. First, this should deal with the new processes that still need to be documented. Once again the approach is to consider the best way of integrating these new processes, and again this should ensure that the newly-integrated and revised data structure will support both the existing process and the new processes.

Once this has been done and there is a complete self-supporting set of documentation, the usual checks should be made to ensure that all the documentation stills holds together. This is no different from the scenarios painted before and the techniques described earlier can be used to make sure that this is the case.

The object of this exercise has been twofold. It was certainly necessary to produce a new version of the model showing the revised, totally integrated structure. It is also important to establish how to get from the old situation to the new structure. The end-result has been defined but the route to this goal should also be charted.

There is now only one final task to perform. There should be a detailed statement of the changes that have been made, not only to the data structure, but also to the processes. This should be done in order to assess the degree of change that needs to be made. This is a statement in logical terms of the changes that have to be made to implement the new model. This form of detailed statement will allow a much more accurate assessment of cost and resource to be made. The exercise has determined the level of change that is necessary and ensured, by carrying out all the checks outlined above, that the route is correct and that nothing has been overlooked. It is now relatively easy to break up the task into individual activities that can be resourced, estimated and controlled at project level.

It should be clear from the foregoing that data analysis can help to control changes to computer systems. Change is one thing, but revolution is quite another! Data analysis is often closely associated with database techniques which themselves represent a massive change in system design philosophy. We will turn our attention to these issues in the next three chapters.

Chapter 18

Going Database

The Story So Far

The basic aims of data analysis are:

- *separation of data and processes;*
- *flexibility and ease of change;*
- *integration and data sharing;*

Database design techniques attempt to preserve these principles in the design and implementation.

So far this book has been very 'conceptual'. There has been little mention of computer systems and systems design as discussion has concentrated on the analytic side of defining and understanding the problems to be solved. This is all well and good. Data analysis is no more than an approach to help the *analysis* of the business and its functions. It does not of itself address the *design* issues associated with actually implementing a solution. The argument is that data analysis will be of great benefit whatever type of solution is adopted. For any such solution, the essential first step is to understand the problem; gaining this insight is totally independent of that eventual solution.

To this extent, this book could stop at this point and ignore the whole issue of systems design entirely. This viewpoint may be logically correct but in reality there is a growing use of certain *design* techniques that follow on naturally from the *analytic* techniques so far discussed. These we will, rather loosely, call database techniques. The guiding principles promoted during analysis are extended to reach down into design to preserve the elegance of the data analysis all the way through to the eventual implementation.

What are these principles you may ask? As a mini-review then, let us refresh our memory of what may well by now be taken for granted.

Firstly, to separate data from the processes that act on that data. Data analysis has led to the development of a data structure that is largely independent of the processing that is carried out. Secondly, flexibility. Some fairly abstract data modelling techniques discussed actually allow the system to respond to changes without a massive re-write of the application. A well developed data model, based on the principles expounded earlier, will insulate the system from change to some extent. This is because the data model is based on fundamental business functions and data rather than any current view of those functions. Thirdly, integration and data sharing. The data model is built around the concept of shared data. In a sense, the data resides in one big pot to which everyone has access. The result is that data is stored in a consistent and economic way and the facility to draw data together across system boundaries is made much easier and reliable.

What is a database? In simple terms a database is any organised collection of data. In this sense the set of files used by any organisation can be referred to as its database, but, in common parlance, a database has come to mean something more than this. It is usual to talk of most file organisations (Indexed Sequential, Random, or whatever) as conventional file structures. When talking about databases, this tends to refer to various specialised forms of file organisation that, although they may use conventional files to hold the data physically, present that data to the outside world in a manner that, to a greater or lesser extent, reflects the structure of the data model upon which the database design is based. Thus there are two main components. First of all there is the database which is the actual data store and is simply a collection of files on one or more storage devices. And secondly, there is the Database Management System [DBMS], this being the software that controls and administers the database allowing the data to be presented in terms of the Data Model. There are a number of different types of DBMS and these are discussed later in the chapter.

Do you need one? There are a number of disadvantages and advantages to taking this route.

Disadvantages

> With current technology, because the data is in effect translated from a physical structure to a logical one, there always is some loss in processing efficiency. If an organisation has processes that are highly time-critical and must be extremely resourceful then it is always more efficient to design a file structure that is closely related to the needs of that process. As technology advances this distinction is getting less important, but it still holds true for some types of process. The concept of data sharing and data consistency also calls for considerable up-front effort in determining the best way of structuring the data for the benefit of the organisation as a whole. Databases cannot be thrown together overnight. They need to evolve as a planned development in order to ensure that the basic design objectives are not compromised in the name of short-term expediency.

Advantages

> Basically this comes back to the main principles again. Processing requirements frequently change and in order to reduce the so-called 'applications backlog', there has to be insulation against change. Because a database structure tends to be process-independent, it is possible to avoid having to change file structures and reorganise data each time the process changes. This reduces the size of the task. The implemented process will have to change, but the data structure supporting the process can often remain unchanged.
>
> The other main advantage is consistency of data. Because all the data is held in a single structure, the ability to examine this structure in different ways is supported in a consistent way. The rise in requests for management information and other 'controlling' rather than processing systems, highlights this need. It is more and more necessary to provide the right quality of information to decision makers in the business. A decision maker has special needs for accessing and using data. These processes defy analysis in the formal sense because each problem requiring a decision will be different from any other. The computer system must support these needs and conventional systems do not readily lend themselves to this approach.

The bottom line to all this must be, "is your company ready for it?" As with data analysis, database developments require commitment from the company if the benefits are to be fully realised. Databases are *not* a panacea for all problems. The system needs careful design and implementation and the pay-off is relatively long-term. If your company needs short-term solutions, go and buy a package, it will get you 80-90 per cent of the way there! We can get some insight on this if we look at the early users of database techniques. Typically, these users were large companies or government agencies and the like, with large volumes of data and the resources and the foresight to go for long-term returns. They were also the types of organisation with a greater need for management information due to the size of the operation. This does not mean that only large organisations can benefit. Rather, these types of organisation have the greatest need for database techniques and therefore entered the field first. The main thing to bear in mind is that data analysis and database techniques do mean a change in philosophy for the systems department. This philosophical change is a necessary precursor to successful database implementation. It is possible to take each systems' data and put each on a separate database. This will help to address some problems, but the fundamental benefits will not be achieved simply by changing the 'file-handler'. Until people are ready to view data as a company resource and are willing to accept the constraints of global accessibility to gain the benefits of this view, database techniques will not yield a full return on the necessary investment.

Given that you still want to proceed with a database and that you feel that the necessary philosophy can be put in place, where do you start? This is dealt with in the next section.

Getting Started

Before beginning design, answer these questions:

- *which DBMS?*
  - *which type?*
  - *which specific product?*
- *what training is needed?*
- *establish the phasing of the system.*

The very first thing to address is the question of commitment. Senior management support will be vital for success. The pros and cons of using database techniques will need to be thought through and their effect in your own organisation assessed. What do you hope to gain? How much resource can you devote to the development? Have you any major time-constraints? These are critical questions which must be answered before you proceed.

Assuming you can obtain the necessary commitment and are happy that a database solution is correct, the next question to answer is 'which DBMS to use'? This question is not going to be answered for the moment. The problem of selecting the appropriate DBMS breaks into two parts. Firstly, which *type* of DBMS to use, and secondly, which specific product of the selected type. The different types of DBMS are discussed in the next section; that discussion should enable the appropriate type to be selected. Once the type of DBMS is determined, this leaves a standard software evaluation exercise to select the best product for the specific needs. The process of software selection for a DBMS is no different from selecting any other software and is outside the scope of this book. Suffice it to say that you will need to draw up a list of the features required and their relative importance and then assess each possible product against those criteria bearing in mind any constraints that may operate on freedom of choice, for example, hardware policy.

For now it will be assumed that you have selected the DBMS you are going to use. Next it is necessary to determine the training requirements. Most vendors of database products supply suitable training programs which may be made available either at the vendor's location or in-house.

Vendors have a vested interest in ensuring a successful implementation of their product and this must include adequate training for the necessary staff. You should, therefore, consult with the vendor to establish how many specialist staff will be required to install and run the project, what level of skills these staff should have, and the amount and nature of specialist training required. It must be said that the amount of specialist knowledge required varies greatly from one DBMS to another and will also be affected by the size of the database project and the nature of the processing. Make full use of the vendors expertise in this area – they should know their product and what back-up is required. Wherever possible try to put the training into effect *immediately* prior to starting development. If the training is implemented too early, the staff will have 'gone off the boil' before they get to use the product. Conversely, if the training is implemented after development has started, staff may have begun to use the product incorrectly or developed 'bad habits' which they will have to un-learn.

The final aspect requiring attention is the phasing of the implementation. If the project is of any reasonable size it is almost certainly impractical to implement the whole area in one go, so it is important to isolate 'chunks' of processing that can be logically implemented in a sequence which causes least user-disruption and yet maximises benefit. The data model and the process documentation is very useful as input to this task. It can be used to ensure, for example, that a process that reads an entity type is not being implemented prior to the implementation of the process that creates it! This may seem obvious – but all the detail needed to ensure a smooth phasing of the implementation is available in the documentation – so use it!

This does not, of course, help to determine the relative priorities for implementation. This can only be done by consultation with user management. Establish what, in their view, would be the most beneficial phasing-in of functions and then check this back against the data model and process documentation to make sure that it is practical. Where necessary, you may have to change the sequencing or move certain functions around to get a feasible timetable. If necessary, do so, but go back to the users and tell them what has been changed and why. Once again, user commitment, understanding and backing are very important and they should be kept informed.

Having decided what is going to be done and how it is to be achieved, it is possible to commence the design work. We must return to that very important question – which database should you use?

Which Database?

There are several types of database product with specific features:

- Hierarchical
    - *ideal for high-performance;*
    - *restricted data structuring;*
    - *difficult to change;*

- Network
    - *good performance possible;*
    - *good capability to reflect complex structures;*
    - *physical data reorganisations can be time-consuming;*

- Relational
    - *natural representation of data;*
    - *insulates data retrieval from physical storage;*
    - *highly flexible;*
    - *slower performance but improving as products mature;*

- Others
    - *Adabas: high performance using indexing;*
    - *hardware solutions: specialist hardware engines to search data files*

There is an ever-burgeoning choice of DBMSs available. A cursory glance at the sales advertisements would lead us to believe that each vendor's product is the very best for all possible applications. This is obviously untrue. Each individual DBMS has various advantages and disadvantages which need to be balanced and assessed to arrive at the best choice for your project in your application.

Start by examining the different *types* of DBMS. It would be unfair of me to comment too rigorously on the capabilities of one product or another, not least because whatever was written would be out of date prior to publication! However, the different types of DBMS are more enduring and it is possible to make some general statements about the capabilities of each. Bear in mind though, that as technology races forward, the distinctions between these different types become more and more blurred.

The basic types that will be considered are Hierarchical, Network and Relational Systems. There is also some discussion on other types of DBMS that either do not fit easily into any of these categories, or which use a novel approach to the problem. The following is not a complete guide to the features of each type of DBMS, but should give the general picture.

Hierarchical Databases.

Chronologically speaking, Hierarchical DBMSs were the earliest form of database and, as a solution to some processing problems, they are still very popular today. The most well-known example of this type of DBMS is the Information Management System (IMS) from IBM. IMS is probably the most widely-used database product in the world, due more to the weight of IBM in the market place and the length of time the product has been available than anything else. As the name implies Hierarchical Databases are restricted, more or less, to a hierarchical view of the world.

In data analysis terms, this means that each entity type can only be at the 'many' end of one relationship and this is obviously a very major restriction on how entity types may be related. There are a number of ways of overcoming some of these problems but each is always some form of compromise and begins to add extra complexity to the structure of the product.

How does a Hierarchical DBMS actually function? This will be described in terms of the simple data model shown in Figure 18.1. Obviously this is a simple structure capable of being mirrored in a Hierarchical DBMS. Figure 18.2. shows a particular example for an IMS implementation. In IMS terms this constitutes one 'record' consisting of several segments. The Root Segment is the entry point into the structure. It contains the key for 'Department A' and any data fields pertinent to a department. The Root Segment also holds a pointer to the location within filestore of the first dependent segment at the 'MAN' level, in this case 'JOE'. This segment, the 'physical child' of Department A, contains the necessary data about 'JOE' and also two separate pointers. Linkage needs to be maintained between all the 'Man' Segments that are related to the same Department. Thus, in this example, 'JOE' will point to 'FRED' who will point to 'TOM'. But 'JOE' is also the 'physical parent' of two JOB Segments, 'J9961' and 'J1246', so a further pointer needs to be maintained within 'JOE'. By the use of such pointers it is possible to move up and down the hierarchy extracting data as required. The terms 'physical parent' and 'physical child' have been deliberately introduced without explanation. In terms of the example described, all segments have been physically related. Each segment may only have one physical parent although a physical parent may have several different types of physical children, and many occurrences of each type of child. Physically related records are normally stored close together in filestore and preferably in the same block, although this is not always possible.

It is also possible to have 'logical parents' and 'logical children' although there are many rules and restrictions governing how these may be structured. It is through the use of these 'logical' relationships that some of the restrictions of the Hierarchical DBMS may be overcome.

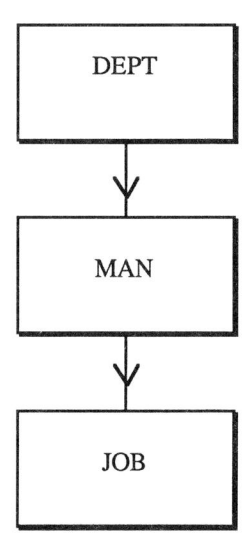

Figure 18.1   A simple hierarchical Data Model

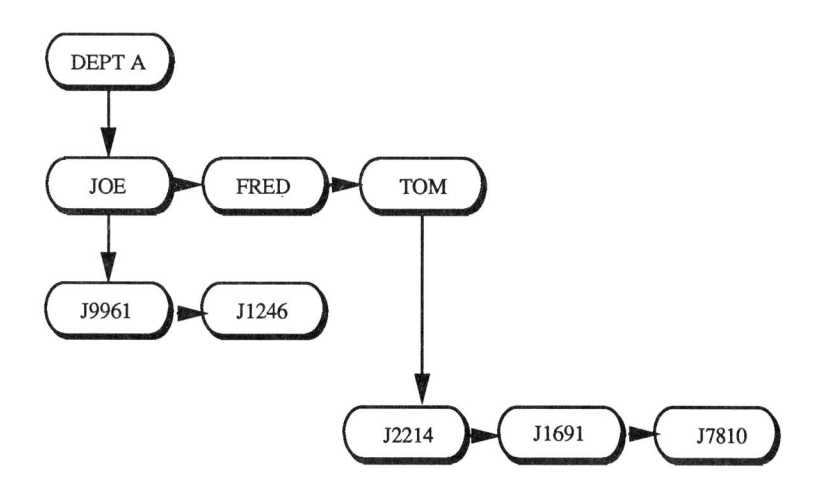

Figure 18.2   A 'record' occurrence within IMS

Going Database

The advantages of Hierarchical DBMSs are normally to be found in terms of performance. The relatively simple structures that can be represented allow for specialised storage and retrieval mechanisms which will enhance performance throughput. Additionally, and at least as far as IMS is concerned, there are many features available within the software itself which allow specific performance improvements to be made if the designer of the database feels that these are required. There are also disadvantages. The simplicity of the structures that can be reflected in a Hierarchical System may require the designer to make many compromises to the original data model in order to obtain a data structure that can be implemented. Additionally, it is not particularly easy to change the structure of the database without considerable specialist involvement.

As a general statement, a Hierarchical DBMS may offer a good solution where the data structure is very simple, where it is unchanging, where the processing requirements are fixed and highly predictable and where high performance is required.

Network DBMSs

Network DBMSs are a more recent development than Hierarchical systems although some of these have been available for about 20 years. The basic standard for Network DBMSs was laid down by the Codasyl committee in the mid-1960s and has basically remained unchanged to date. A Network DEMS should be able to reflect accurately any Data Model produced by the techniques outlined in this book.

The major concepts within a Network DBMS are those of Owner, Member and Set.

In terms of the data model in Figure 18.1. each Department would be the Owner of any number of Man records. Each Man record would be a Member of a Set of records owned by one Department record. The normal diagrammatic convention for describing an example of a set occurrence is shown in Figure 18.3. for the data occurrences used earlier. Therefore, Department A owns a 'set' of three Man records and these are linked together by means of a closed loop of pointers. Similarly, 'JOE' has a set containing two JOB records which once more are related *via* a closed loop of pointers. This is fairly similar in principle to a hierarchical DBMS. The differences start to occur when more complex data structures are required. In Network terms, a record may be a member of as many sets as required, provided that each is of a different set type. Thus, 'JOE' can only be owned by one Department at a time, but he could also be a member of a set owned by any other type of record. This may become clearer by taking another example. A more complex data model is presented in Figure 18.4. This type of situation can start to cause problems for Hierarchical systems as the 'MAN' entity type is the subject of two relationships, one from Department

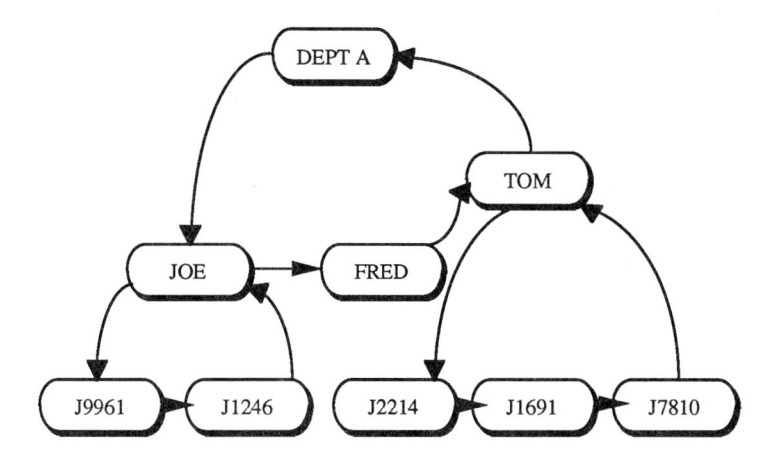

Figure 18.3   A Network DBMS structure

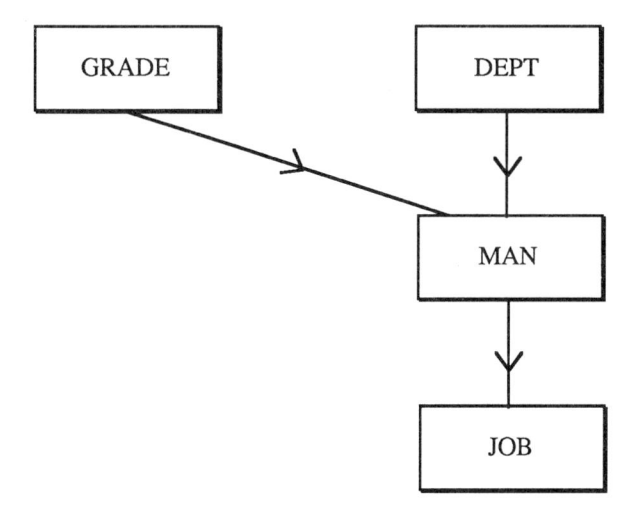

Figure 18.4   A more complex Data Model

Going Database

and the other from Grade. In Figure 18.5, some sample data occurrences are shown for this structure. Also shown are the pointers that would be maintained by a Network DBMS. The significant point here is that the Network DBMS allows a record to be a member of multiple sets so 'BARRY' is owned by Department A and GRD1. It is therefore possible to describe fairly complex structures within a network DBMS.

The main advantages of Network systems are performance oriented. Due to the use of physical pointers, navigation around the structure can be very efficient and most Network DBMSs provide a variety of design options for increasing performance. Network DBMSs can also handle flexible data structures, but only if these have been carefully pre-defined.

The major disadvantage resides in the use of physical addressing, and the allied fact that new routes through the data cannot be efficiently exploited without reorganising the stored data.

In general terms, a Network DBMS will provide high transaction throughput where the processing requirements are clearly defined. Responsiveness to *ad-hoc* enquiry is often limited and reorganisation of the database can be a costly overhead.

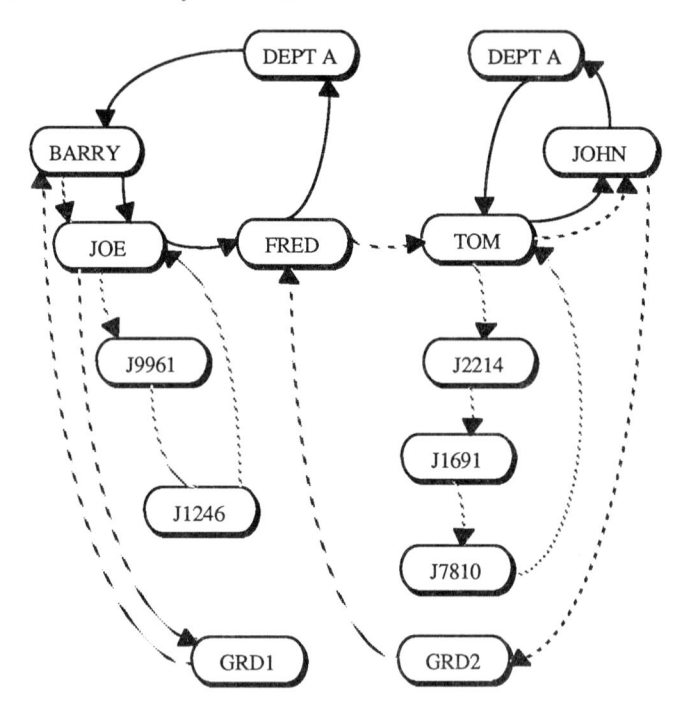

Figure 18.5   Some sample occurrences for the preceding Data Model

Relational DBMS

In recent years, a new type of database has become increasingly popular. This is the Relational DBMS. Relational systems represent data as relations or 'tables' in a manner which is very similar to the results of data analysis. The relational representation of Figure 18.5 is shown in Figure 18.6. Notice that no explicit relationships are defined within a relational DBMS. The 'relationships' are logically dictated by common data items between tables.

DEPT Relation

Dept. No.	Name
A	Machine Shop
B	Despatch
etc.	

GRADE Relation

Grade No.	Name
1	Unskilled
2	Semi-skilled
etc	

MAN Relation

Name	Dept. No.	Grade No.	Age	Address
BARRY	A	1	46	24 Station Road
JOE	A	1	49	46 Acacia Avenue
FRED	A	2	31	12 The Avenue
TOM	B	2	37	12 High Street
JOHN	B	2	28	68 London Road
etc				

Job Relation

Job ID	Man	Date	Bonus Rate	Hours etc.
J9961	JOE	12.1.86	1.5	10
J1246	JOE	15.2.86	2.0	6
J2214	TOM	19.7.85	2.0	7
J1691	TOM	21.6.86	1.0	4
J7810	TOM	01.10.85	1.25	15
etc				

Figure 18.6 Basic relational tables

Thus the Department Number item occurs in both the Department relation and the Man relation indicating that a Man can only 'belong' to one Department. Relational DBMSs are explicitly designed to recognise this commonality of attribute and to enable data navigation to proceed on this basis. As data navigation is carried out on the content of the records rather than physical pointers, the records can be moved around in filestore or even redefined with different data without requiring massive changes to the structure or definition of the database.

The major advantage of this is in terms of sheer flexibility. With relational systems it is much easier to change the way one wishes to look at the data, and normally this can be accomplished with no physical change to the data. Also, the physical storage can be completely reorganised without the logical structure changing and this can also be a transparent activity.

The main disadvantage is in terms of speed of processing. To be true to its theory, a relational database should have no features that are dependent on physical data storage. The result of this is that where data is to be retrieved, no physical attributes, such as pointers, should be used and therefore data retrieval can be slow in some cases. In actuality, all relational systems available today use some form of physical addressing to speed up retrieval in some situations. This is still essential, especially where data volumes are large.

In general, a relational database should be considered where issues of flexibility outweigh those of performance. It should also be noted that the power of relational systems is improving as the products become more mature and as hardware becomes cheaper and faster. Therefore, the range of applications for which relational solutions may be deemed suitable is gradually increasing.

Other systems

There are some other types of DBMS which may also be effective in solving a particular set of problems. One of the most widely used of these products is Adabas which is not really a network or a relational product in the true sense. Adabas relies upon flexible and powerful indexing strategies and has been shown to be capable of high performance and yet still retains considerable flexibility.

Also worthy of mention are Database Machines such as the Britton-Lee Intelligent Database Machine (IDM), Teradata and similarly, ICL's Content Addressable File Store (CAFS) processor. Regardless of the type of DBMS that these products support, their common theme is the use of specialised hardware techniques to provide very fast data searching capability. With a true relational system all retrieval must be based on data values, and this implies the availability of an efficient search engine to access this data. Database machines are a potentially revolutionary solution to certain problems. Whether the products will live up to their expectations only time will tell!

It is obvious from the foregoing that performance criteria are highly important in determining which type of DBMS to use, to say nothing of which actual product. The main performance-related issues are the throughput required of the system, the flexibility required, the basic complexity of the different types of process to be supported, and also the complexity of the data structure. The processing profile to be satisfied is essential input to the selection process, and how well each of the products satisfies this profile should be measured.

There are also a number of other factors that will need to be considered. What type of application is being addressed? For high-throughput production systems then a Network or Hierarchical system may be suitable. If the primary concern is with flexible and responsive reporting and enquiry facilities, this is more the province of Relational systems. In truth, of course, the application is usually a mixture of these areas and the choice is less well-defined. Also there may be hardware considerations that in some way preclude or restrict your choice. If you are in a committed IBM shop and you know that no other type of hardware will be countenanced, there is little point in considering products that will not run in an IBM environment!

In addition to the basic facilities required for the application it is also useful to consider how important various back-up facilities will be to this particular solution. For example, what are the security and recovery requirements and can these be satisfied by the software? Can recovery be completed within a time-frame acceptable to the user and what are the implications on the business of a system failure? What other database facilities are available and are these important to your application? The ability to load and unload data to or from the database may be very important. Similarly, reorganisation or restructuring capability might figure highly in terms of system priority.

The last set of factors to be considered are any additional software features that might be available and their importance should be assessed. Such extras as Data Dictionaries, Report Generators, and well-developed Query Languages might be very useful. It is increasingly common for the total development environment to be closely integrated around a central database product. Selecting a database might, in effect, change the whole development approach within the organisation making increasing use of Fourth-Generation Languages (4GLs) and Application Generators.

In conclusion then, 'which database?' is an involved question which should be given careful consideration from many different angles before coming to a conclusion.

Phasing the Implementation

Whenever possible implement a small self-contained area first to prove the approach used in a low-risk way. If this cannot be achieved then take particular care over the following points:

- *set out a clear approach;*
- *produce draft standards and adhere to them;*
- *set out guidance notes on new techniques;*
- *logical database design;*
- *physical database design;*
- *process code production;*
- *implementation;*
- *review the whole exercise post-implementation;*
- *database tuning.*

Almost as important a question as "which database?", is that of determining which parts of the data model are to be implemented using this new database software. There are two ways of approaching this problem: the first is the more idealistic approach, the second is what tends to happen in reality!

The ideal text-book approach is to take a small fairly self-contained area of the business, and gently and gradually implement a system to service this business area using these new database techniques to the full. With this sort of approach there is time to get to know the software and the various design options. There may even be sufficient resource to enable the testing of different solutions and options before making a design decision.

If it is possible to take such an approach then do so! Once the area has been determined, and presuming that modelling has already been completed, then design can begin. There will be many design issues which are unchanged by the underlying database (such as screen design, report definition), but the database itself will need to be designed and the use of 4-GL's and other application generators will affect the design process. Application generators tend to affect the design in a product-specific way and they are outside the scope of this text. The major database design decisions are covered in the remainder of this chapter with other design issues in Chapter 19.

Once the database system has been designed then the process code can be written and the testing of the developing application initiated. This function is not significantly affected by the use of database techniques and the traditional techniques of clear programming and thorough testing are no less relevant than they have ever been.

Similarly, the implementation phase still retains the same purpose. Any user-training in the new system will need to be planned and executed prior to live running and the change-over to the new system will need to be established with all due regard to the users' operational requirements.

Once the system has been successfully implemented, it is time to sit back and review the whole process and use the practical experience gained as basic input for setting up standards and guidelines for future projects. This should pay equal attention to things that went wrong as well as things that went smoothly. Did something go wrong because a technique was flawed or merely because it was used incorrectly? A thorough and critical review will reveal much about how future projects should be controlled. Full use of this information should be made and the lessons worked into the standards.

The foregoing is very much an ideal and, although it may occur occasionally, the reality is usually very different!!

The decision by an organisation to utilise database techniques often signifies a major investment by that organisation. New people-skills have to be learnt, new software needs to be purchased and, in many cases, additional or replacement hardware also figures in the equation. When the level of investment is high, there is great pressure to realise some significant return on the investment in a reasonable period of time. The result of all this is that the first database application to be attempted is in a large, complex area of the business and, more often than not, this application area also has high visibility throughout the rest of the company.

Let us not get drawn into the argument of which is the better approach. Being 'dropped in at the deep-end' can work very well and produce an excellent end-result, but such an approach does require more careful planning and the exercise must be split down into a greater number of tasks.

First of all, there will normally be more people involved with the project and timescales will often be tighter. Therefore, the approach needs to be set out clearly before the project starts and, once work has begun, it should not be changed significantly unless fundamental problems are encountered. Draft standards should be determined and produced to govern the techniques used within the project. Educated guesses will need to be made about the best techniques to use and be set out in draft form. Project management should ensure that the standards are then adhered to except under exceptional circumstances.

The design phase itself can also be split into several sub-tasks. First there is what may be termed 'logical database design' which determines what record types will be used, and what data items are in each record type. It will also consider what indexing strategies are needed and, where appropriate, the use of such things as pointers or foreign keys.

Once the logical database design is complete, two other major activities may be fired off. The first is the physical design of a test database. How large should it be? What is the relationship between different record types and the underlying physical files? The physical design will also consider security and recovery strategies as required. This is an important issue for a production database, but for a test version the whole issue of physical design is less critical unless performance benchmarking is to be an important part of the test system.

In parallel with the physical design of the test database, the development of process code and initial testing may begin. Although the test database will have to be available for program testing the physical design issues do not normally affect the program writing phase since the programs are interested in the logical, rather than the physical, structure of the data. The two tasks can, therefore, be significantly overlapped thereby saving on overall development timescales.

Once the test database is available and testing is underway, it is necessary to concentrate on the highly technical task of live database physical design. Depending on the DBMS chosen, there are often many factors that can significantly affect performance of the live system and full and careful consideration of these issues is essential. Some of the more fundamental of these issues are described in the next chapter. In most cases, the vendor of the database software will be able to assist in these areas, especially as specific design trade-offs are often only relevant to a specific product. If you are inexperienced in the software, it is as well to seek advice here. The rules that apply to one DBMS may have totally different effects when applied to another!

Once the production database has been designed and the programs have been fully tested, the new system can be implemented using standard hand-over procedures such as parallel running to ensure the integrity of the new system. Then all that needs to be done is to review the whole exercise!

This review needs to address three specific objectives if a major implementation has been tackled as the first system.

Firstly, the standards that were 'invented' at the start of the project will need to be redrafted. Practical do's and don'ts will have been learned through using the software and these can be incorporated into the standards. At the same time, any standards found to have been unhelpful should be removed.

The second objective is to re-evaluate and improve the project control techniques for database projects. Were adequate time and people resources allowed when the project was planned? Were any activities over-estimated? A major development activity will give lots of feedback which can be used to help achieve better project control next time.

Finally, but very importantly, the implemented database will need to be tuned. The physical design will have been based on *predicted* loadings for the system. The *actual* system loading may be quite different in certain key areas. Many DBMSs have a number of performance aids built into them. These need to be monitored and appropriate corrective action taken as required. This task must continue throughout the life of a system as usage patterns can easily change. In particular, frequent performance monitoring is essential for the first year of live running while the usage of system settles down and any initial performance problems are ironed out.

Database Design

Although the sequence is not too important, the following have to be addressed:

- *translating the data model into a database design;*
- *performance issues;*
- *sizing;*
- *security and recovery;*
- *redundancy and housekeeping.*

The next few sections discuss some of the major points within the task of database design, but there are some general points that should be made before turning to specifics.

First, the following sections do not present a full discussion of database design techniques. Volumes can, and have, been written on how this should be approached. It is the aim of this chapter to give a few pointers to the major decisions that have to be taken. The relative importance of each issue will vary from one DBMS to another as well as from one application to another, but these major decisions will still have to be made at some point and the following few sections describe the general issues surrounding each one.

The remainder of this chapter describes certain techniques for translating the data model concepts into a database structure. Chapter 21 addresses various technical aspects of database design. These are performance issues and also sizing considerations for physical store. Security and recovery techniques are also discussed. Chapter 21 also examines various design considerations which tend to affect the later stages of the project – sometimes well after the initial implementation date. These issues include built-in redundancy and consideration of data reorganisation and housekeeping strategies.

The sequence of these issues and their attendant design decisions is not inviolate and can be re-sequenced to meet the needs of the project. However, it is rare that any of these points can be safely ignored within the design process. In other words, it does not really matter in what sequence these things are done – providing they are done!

Entities vs Records

> *Entity types should only be split across database records, or combined into records where there are quantifiable benefits. For example:*
>
> * *combining entity types: the only time this is likely to be justifiable is where several entity types have very similar attributes and are processed by the same or similar processes;*
> * *splitting entity types: in certain special situations this may be beneficial:*
>   * *where sub-sets of attributes are processed independently;*
>   * *to enforce security control;*
>   * *to allow 'summary' records of history to be held;*
>   * *to include selected derived data.*

Having expended a lot of time and effort in determining the entity types and fully documenting these within the data model, the data structures that will have been built using the data analysis techniques will be relatively simple to implement on either Network or Relational DBMSs. Many of the structures could also be implemented using a Hierarchical DMBS, although the compromises that would have to be made would be more fundamental.

In order to come up with a straightforward first-cut database design which will work, it is possible simply to define each entity type in the data model as a record type within the database. This will at least preserve the original elegance of the data model and also ensure that any changes made to the logical data structure are made for good reasons.

In 90 per cent of cases this one-for-one conversion between entity types and records will be quite adequate. The other 10 per cent of occasions do require some further actions. Basically, given that the data in these remaining entity types should appear somewhere within the database, one of two actions can be taken: either several entity types may be combined into one implemented record, or one entity type may be split into several records for implementation.

The first of these to consider, and the more likely, is combining several entity types into one record. The commonest situation here is that there are several entity types with quite similar attributes and, when examining the processing of these entity types, it is found they are used in the same way by similar processes. Although the entity types may be logically distinct, this recognises that from a processing point of view they are effectively the same and, therefore, the same implementation structure can be used to hold them.

Even in these circumstances, entity types should not be combined unless there is some quantifiable benefit. There may be some critical performance aspect that is eased by such a measure but, if no significant benefits are to be gained, the entity types may as well be implemented as separate record types as this ensures greater flexibility.

There may also be some cases where there is a need to split up an entity type into several records. This will be rare if the techniques in earlier chapters have been used to their full extent as these techniques themselves tend to produce fairly compact entity types with only a few attributes in each one. In practice it is only reasonable to split up entity types that have a relatively large number of attributes. There are two reasons why this might be done. Firstly, not all the attributes of the entity type may be accessed by the same processes. This should have already been documented during the resolution of the results of process and data analysis. If it is found that only sub-sets of the attributes are used for a process, and also that these sub-sets are not intermixed across processes, then there may be good reason to split the entity type into separate record types. This may enable some data to be kept in one physical file (perhaps for on-line use) and the rest of the data kept on off-line storage – for enquiry overnight or whatever. This may be seen to give a number of distinct advantages for performance or other reasons. Again these techniques should not be implemented for the sake of it. If you cannot show a quantifiable benefit for your system if you take this action, then don't do it!

There is a second reason for turning an entity type into several record types. This is done, not on the basis of attributes, but by dividing the entity occurrences into several groups. In a stock management application, for example, there may be a data model similar to Figure 18.7. Now let us assume that for some high security reason, users from each warehouse will have their own view of the system so that they can only see details of stock being held at their warehouse.

By splitting the cross-reference entity type into a number of separate records distinguished by warehouse code, the processes to enforce this security may be greatly simplified, whilst still allowing someone with sufficient authority, say head office, to obtain the full picture. Physical data security could also be enforced by such techniques, as each distinct record type could be targetted to a different disc drive if the DBMS has this capability. However, it should also be noted that the foregoing is not the only way of achieving this result and the pros and cons of introducing what are, in effect, artificial data types into the implemented system should be carefully assessed. After all, you have just spent much time and energy using data analysis to remove all such artificial data!!

There is one final situation where record type design may depart from the entity types already documented. In the data analysis phase an infinite history of data was assumed. Although in logical terms this may be useful, in practical terms it gives two distinct difficulties. First of all, it is unlikely that there would be sufficient on-line storage to hold all occurrences of all records for an infinite period. It will be necessary to archive historical data from time to time. Housekeeping and archiving strategies will be discussed in the next chapter. For now it will be assumed that data will need to be archived off-line and that therefore some form of summary record of the archived situation will need to be kept. This summary record will not appear in the data model, but will need to be designed into the database.

The second situation which will cause 'invention' of record types is concerned with performance. There will be many situations where derived data is needed as output and in some cases it would clearly be inefficient to have to calculate this data from source each time.

Where the same derived data is frequently required and constant recalculation of this would cause a performance problem, this derived data should be designed into the database, and relevant procedures put in place to maintain it to the required degree of accuracy. Although derived data should normally not be included within the data model, there will often be situations where it is clearly advantageous to hold this data explicitly within the database.

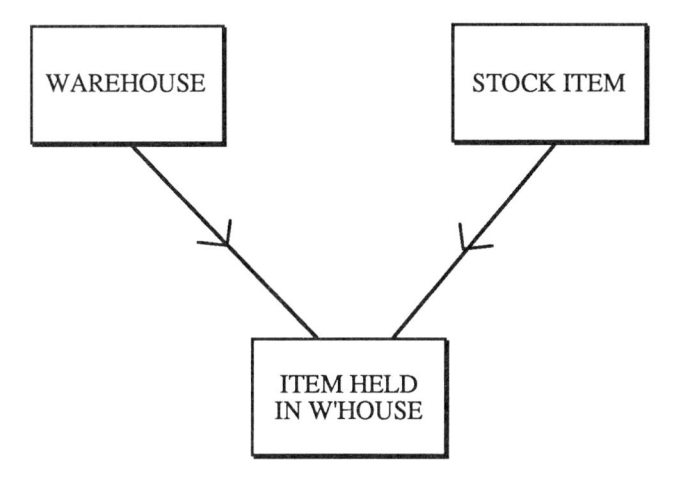

Figure 18.7   Extract from a stock management model

Attributes vs Fields

Turning attributes into data fields is normally straightforward but beware:

- *are all Identifying Attributes really needed?*
- *examine the implementation of Supporting Attributes if a non-relational implementation is used.*

As with entity types, there is a broad analogy between the attribute at the data modelling level, and the data field within an implemented record. It will normally be the case that nearly all the attributes for an entity type will appear as data fields in the appropriate database record.

Of course there are exceptions! Throughout the logical modelling it was maintained that each occurrence of an entity type should be uniquely identifiable from every other occurrence of that entity type by some predictable combination of attribute values. Many database systems do not insist that this be so and, in fact, there is often special processing specifically included to deal with duplicates. Many theoretical puritans would argue that a unique key is absolutely essential to correct design. I am not so certain of this. Unique keys certainly simplify many processing tasks, but in some cases the lists of attributes defined as Identifying are so long-winded, or in other ways very esoteric, that one wonders whether there is much to be gained by insisting on this uniqueness. The acid test is really concerned with how each record is to be accessed. Is access to the record effectively random (" Get me Customer X"), or is the access primarily via another record (" Get me all locations of Customer X")? If there is good reason to believe that unique access to occurrences of this record type is extremely unlikely, it is reasonable to query the necessity of implementing all the Identifying attributes. Again, this is a compromise of the original design. This action should not be taken without a known benefit, but if it can be justified – why not?

The other main point about attributes concerns those that are there in a purely Supportive role. In relational databases, relationships between the various record types are defined by the commonality of attributes. Supporting attributes are therefore absolutely essential for the correct operation of such a system and they must be explicitly implemented if the relationship is to be used to access data.

In Network and Hierarchical systems, though, relationships are effectively maintained by internal physical pointers from one record to another. These pointers are used by the DBMS to control navigation through the database. So for these systems, carrying additional attributes purely to support relationships which are already supported by internal pointers is clearly an overhead. Before getting out the red pen and deleting them from the implementation, though, are you sure that the attribute is not, *and will not*, be used for some other purpose? Is it a significant overhead anyway? If you later migrate from one DBMS to another, the task will be eased if the implementation mirrors the data model as closely as possible.

Relationships vs Sets and Indexes

For each relationship the following questions need to be answered:

- *is the relationship needed to retrieve data?*
- *what is the cardinality of the relationship?*
- *should the relationship be ordered?*

The answers to the first question should be available from the results of the Process analysis and the second should be answered by data analysis. It is only for the third question that further research is needed. Details of the type of processing using the relationship should reveal the requirements in this area.

The final aspect of the data model which needs to be considered for implementation is that of relationships. There are several ways of implementing a relationship within a database. The choice of method is somewhat dependent on the DBMS, but not exclusively. The first question to ask is "should the relationship be implemented?" The relationship should be made explicit *via* some form of set or index if there is a need to retrieve data *via* the relationship. If the relationship is not needed for data navigation the requirements can probably be satisfied by including appropriate attributes in each record type.

For example, consider Figures 18.8 and 18.9. Figure 18.8 shows what is sometimes called a high-cardinality relationship. Cardinality is the term used to describe the number of occurrences at the 'many' end of a relationship. In Figure 18.8 there will obviously be a large number of POLICIES for each UNDERWRITER. If this relationship were implemented explicitly, *via* either pointers in a network system or an index, then it would be possible to retrieve POLICIES by UNDERWRITER with ease. Would this be particularly useful? Normally, in order to retrieve that much data it would be more efficient to retrieve all POLICIES sequentially (thus maximising I/O efficiency) and then sort them into the required sequence. Remember 'Underwriter Id' would be a supporting attribute within POLICY.

Figure 18.9. shows the opposite effect. The relationship between VEHICLE and POLICY would have low-cardinality as only a small number of VEHICLES would be related to each POLICY. Given that there would be frequent need to retrieve all the VEHICLES for a POLICY with ease, access to these precise VEHICLES out of the several thousand or so occurrences that might be stored would need to be efficient. In this situation a direct means of retrieving all VEHICLES related to a POLICY, and only these

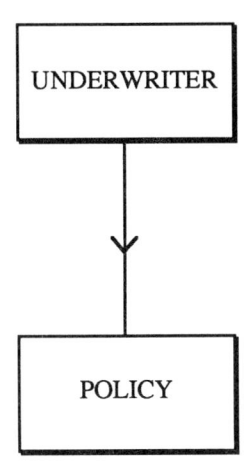

Figure 18.8 A "high cardinality" relationship

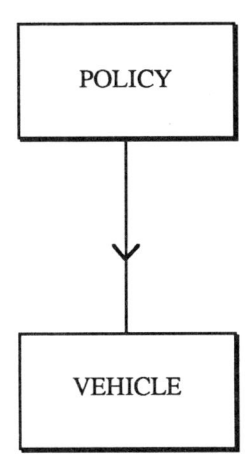

Figure 18.9 A "low-cardinality" relationship

VEHICLES, would be very helpful. In this situation pointers or indexes would be very useful.

As can be seen, relationship cardinality is an important factor in database design and this, allied with the use to be made of the relationship, makes a significant contribution to the final database structure.

Having determined whether the relationship should be implemented, the next important decision concerns ordering within the relationship. With both indexes and with Network Database sets, it is possible to store occurrences in some sequence based upon certain attribute values. Once again the type of access to be made is important. Storing record occurrences in sequence will always cause some additional overhead and in some cases sequenced sets or indexes can degrade performance unacceptably. In general terms, sequencing can be useful in some situations and unacceptable in others. As an example, look at Figure 18.10. This shows two relationships, both with reasonably high- cardinality. It may be necessary to provide some form of 'browse' facility on MAKES and MODELS, allowing the terminal user to search for and select the required MODEL. In this case ordering the relationship – probably alphabetically within MAKE – would enable this facility to be more easily implemented. It is also likely that the MAKE/MODEL relationship is relatively stable and would not be updated too frequently. In this case, the retrieval advantage would probably outweigh the storage overhead and ordering would be useful.

By contrast the relationship between MODEL and VEHICLE is very different. It would be very nice to be able to retrieve all VEHICLES of a given model in some order, but would it be useful? Often VEHICLES would be stored and retrieved randomly and it would incur a substantial overhead if VEHICLES had to be stored in sequence within MODEL. The pros and cons need to be carefully assessed.

The basic 'rules' governing translation of the data model into a database design have now been considered. It is now time to move onto other relevant factors.

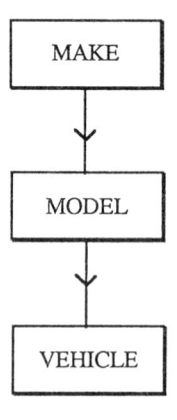

Figure 18.10 Good and bad examples for 'ordered' relationships

Chapter 19
Physical Design Issues

Performance

Specific design issues that may affect performance include:

- *the number of disc accesses;*
- *the pattern of access, that is, random or sequential;*
- *the effects of block size and record 'clustering';*
- *update overheads for index or pointer maintenance;*
- *disc and record contention;*
- *double update possibilities.*

There are many factors that can affect the overall performance of a database system. This section will only scratch the surface of the major issues, but there is one general principle to be remembered. Except in particular situations, database systems will not give 'efficient' performance in the traditional sense. The essential feature of database systems, flexibility, is contrary to efficiency. There is a design trade-off here and it can make these factors more relevant. If you desire the flexibility of a database system and still require high efficiency, then performance factors must be carefully measured and controlled.

Disc access

The major factor in performance evaluation is the number of disc accesses. The more fragmented the data – a natural result of data analysis – the more physical accesses will be needed. The number of accesses is a good predictor of response time and if this is unacceptable to meet design requirements then action will need to be taken. Many DBMSs provide features which will help in minimising accesses and these should be employed as necessary. In other cases it will be necessary to 'compromise' the data structure to improve performance. This is usually achieved by moving away from the elegant logical data structure to something less logically clear. This may improve performance but it also reduces flexibility.

However, it is not just the number of disc accesses that is important but also the pattern of those accesses. Sequential reading of physical disc blocks will be much less costly than random access. This means that if the amount of random retrieval can be reduced, by placing data more efficiently, then overall performance will be improved.

Block size

Allied to disc access is the effect of block size. Where facilities exist to 'cluster' related records together on physical storage, the block size can make a difference to the efficiency of the system. A small block may contain perhaps ten related records, a large block perhaps fifty. In disc access terms, the overhead of reading the larger block is minimal as the seek time is far greater than the time taken actually to transfer data into store. So, if records are being clustered together, the block size should be chosen so that the clustered records can physically reside on the same block. Where there is a high volume of sequential disc accessing, use of larger block sizes will reduce the number of reads necessary and this again can improve efficiency. However, where access is essentially random, a smaller block size will reduce contention and improve efficiency as less data will be transferred to and from store. But remember, not all DBMSs have data clustering facilities, and some do not allow a choice of block size.

Updates

There are also update overheads to consider when judging performance. This is quite straightforward to estimate. When a record is stored, the number of indexes addressing that record, and where relevant, the number of Network Database sets in which the record participates, will grossly affect performance. Where pointer-based sets are involved, the records on either side of the new record may also need to be updated to maintain the integrity of the set. For indexes, the index blocks will require updating and possibly re-balancing. All these extra updates may cause additional physical accesses and these need to be accounted for in any sizing exercise.

Contention

The next area to be considered is that of contention for physical resources. This occurs both at the disc and record level. For discs it is necessary to estimate the number of accesses being directed to a specific disc-drive during a typical mix of transactions. If the number of requests begins to exceed the capability of the disc-drive to service those requests then queues will start to build rapidly and performance will degrade.

The system should be designed so that the spread of accesses per drive is reasonably well-balanced and within the capabilities of the drive itself. If not, the spread of data across drives will have to be adjusted.

Contention can also occur at the record level. In most systems, there will be a small number of record occurrences that are accessed with very high frequency. These are, in general, commonly-used codes or other reference data. Where system throughput is high, record locking can cause queues to build up for requests to read those specific records. In these situations, overall disc accesses may look acceptable, but performance will still degrade at peak load. A common solution to this problem is either to lock this data into core to speed access, or to make several copies of the record available and balance accesses across these copies. This latter solution normally entails compromising the data structure somewhat to permit several copies of the records to co-exist, normally distinguished by some additional attribute which can be used to control access to the different copies.

Double updating

It is also possible to get problems in terms of double updating of records. Although all respectable DBMSs will prevent simultaneous double updating, it is possible to encounter problems in this area. Typically this will occur in the 'Airline Reservation' scenario – see Figure 19.1. In this scenario, the record locks are released after enquiry to allow other access as no update has been made. When two terminals try to book the same seats then there is a problem of which the DBMS will not be aware. The only way to prevent this problem without unnecessary locking, is for the process code to check explicitly that no other updates have occured between enquiry and update. For large, multi-user systems this factor needs careful judgement, but the problems are avoidable.

1	Terminal A reads Flight B991 details for enquiry
2	Terminal A informed that 2 seats remain and locks released
3	Terminal B reads Flight B991 details for enquiry
4	Terminal B informed that 2 seats remain and locks released
5	Terminal A now 'books' the 2 seats - record locked, updated and released
6	Terminal B now tries to book the 2 seats - unless availability is re-checked by process code, the flight will be overbooked

Figure 19.1   Sequence of events resulting in double update problem

When estimating performance it is necessary to compile two sets of detail. First of all, the disc access pattern for each process needs to be calculated. Secondly, the aggregate access pattern for typical process mixes should be calculated. The results should point up any major problems and action can be taken.

It is also useful to distinguish between the performance factors affecting on-line usage and those concerned with batch-type processing. Typically, it is necessary to apply more attention to on-line response and only to tune batch performance where this is absolutely critical. Beware, if any batch processing needs to be carried out whilst on-line access is also available, then the degrading effects of this batch processing will need to be considered.

In all cases though, the factors affecting performance will vary greatly in importance from one DBMS to another and each vendor will be fully aware of the relevance of each factor to his own product. Always consult with the vendor fully when measuring performance and make full use of his knowledge and expertise!

Security and Recovery

Some major factors include:

- *speed of recovery and consequences of down-time;*
- *available utilities and their sophistication;*
- *on-line/off-line security copies required?*
- *rollback facilities;*
- *rollforward facilities.*

It is usually necessary to formulate a security and recovery strategy as a part of database design. Where a central shared database is being designed, this issue is very important and a well-thought-out strategy can save many problems if difficulties should arise.

There are a number of factors to consider when planning recovery. The first set is concerned with security performance. First of all, how long is the on-line day? The amount of off-line processing time available will determine how much security is possible. If the on-line day is long and the database large, the amount of time for traditional file security may preclude securing the entire database every day. Secondly, what are the permissible recovery times? If the system crashes, the user will want recovery to be effected more or less quickly. Where quick recovery is required, the security 'checkpointing' must be frequent to ensure that only a small amount of work needs to be recovered. Where a system is primarily for enquiry only and none too critical, a longer recovery period may be inconvenient but acceptable. This is allied to the third point: the consequences of a failure in a front-line trading system may cost a considerable amount in lost business. Loss of an MIS system may incur the wrath of senior management but not affect day-to-day trading. By assessing these three factors, the designer will have a clear idea of how quick recovery needs to be and how much resource should be allocated to providing a slick and effective recovery strategy.

Normally it will be necessary to take security copies of all or part of the database with reasonable frequency and there are three ways to approach this. The most common is to make off-line copies of the database at the end of the day. These copies may either use specialised utilities which are less efficient to run but provide useful features, or simply by using operating system specific file or disc copies. These are normally highly efficient although somewhat basic in functionality. All DBMSs provide some form of off-line security dump facility.

In some more sophisticated systems there are facilities for taking security dumps while the on-line service is running. In many cases these utilities can only secure parts of the database not available for update. Where the on-line day is long, or frequent dumping is required, these utilities can be usefully employed.

The most sophisticated form of security dumping allows on-line dumping to occur whilst the database is available for update. The controls for ensuring consistent recovery from this type of facility, have to be highly sophisticated as the security copy itself may show an inconsistent picture of the database. This type of facility is only available in some of the very advanced DBMSs on the market. If high security and high availability are required, this type of facility will give considerable extra resilience.

Any 'Rollforward' and 'Rollback' facilities must also be considered. When the system is restarted after a system crash or disc error, it will be necessary for any transactions incomplete at the time of the error to be backed out and any updates that may have been made should be reversed. This is Rollback and is available in one form or another with most multi-user systems. The precise workings of the Rollback facility will be important in designing correct restart procedures and are DBMS-dependent.

If a major error occurs and data needs to be restored from the security copies, it is also very useful to be able to reapply any updates to the database that were successfully completed after the last dump but before the point of failure. This is the Rollforward capability. It is normally achieved by some form of transaction log that maintains a record of all database updates and makes these updates available for recovery purposes. Although many database products have some form of Rollforward ability, there are major differences in their way of working and efficiency. It will be necessary to consider the implications of these facilities when designing security.

All the foregoing factors need to be carefully weighed when formulating a recovery strategy and the best combinations of features will vary from one application to another and from one installation to another. Again, the best advice is to consult the DBMS vendor who has a vested interest in ensuring that the implementation is successful. Their detailed technical advice is normally freely available.

Storage

The options available for manipulating physical storage will depend on the DBMS used. Features to consider are:

- *the spread of data across disc drives: the fewer the drives the higher the risk of contention;*
- *the placement of data across discs to optimise and simplify data security and recovery;*
- *use of 'mirrored' discs;*
- *locking frequently-accessed data into core;*
- *do not forget about growth.*

When designing a database system it is also useful to consider the issue of physical storage requirements in its own right. It is true to say that many storage considerations also affect both performance and security but it is beneficial to pull all the threads together in this section.

First of all, performance considerations. There is an immediate trade-off between the number of disc drives and performance. If the database is spread over a large number of drives, the number of accesses to each will be lowered and the likelihood of running into contention is reduced. On the other hand, the fewer disc-drives used, the lower the cost of the implemented system. This is a subjective trade-off. It is relatively easy to calculate the minimum storage resource required to allow the system to operate but how much additional resource can be allocated is a cost which needs to be assessed.

Secondly, security and recovery strategies can be affected by the layout of physical storage. The effects of a disc crash will be determined by what types of data, and how much, is stored on that device. In some cases it may be possible to operate the system in degraded mode if a disc goes off-line. If this capacity is important then it is worth placing the data on physical storage so that it can be made possible. Another allied factor is concerned with security archiving strategies. If file or disc utilities are being used to make these copies, especially if the whole database is not being secured at the same time, then the unit of archiving, disc or file, should be relatively self-contained from the data point of view. This is important. When recovery has to be made the unit of recovery will be based on the unit of security. If disc archives are used then recovery will take place on the basis of whole discs.

By postulating likely failures it is possible to calculate what would need to be recovered and therefore how long it would take. By adjusting the storage layout it is sometimes possible to reduce the recovery down-time and improve system resilience.

Some high-tolerance systems cannot afford to be degraded by the effects of a disc failure. In some DBMSs this can be virtually assured, at a cost, by using 'duplexed' or 'mirrored' discs. In these systems, two separate copies of the data are held on different devices and the system handles the additional overhead of updating both copies of the data whenever an update occurs. For retrieving data, the accesses to each copy are often balanced to minimise disc contention and this by itself may solve some contention problems. In the event of a failure of one disc, the system will normally continue to run in 'simplex' mode accessing only the good copy of data. Once the failed device is returned to service it is normally automatically updated by the system to bring both copies back into line. The additional costs result from doubling the amount of disc space required and also incurring twice the number of physical writes whenever an update occurs. However, the gain in resilience can be significant, some contention problems may be eased, and the frequency of security archiving can normally be reduced as a direct result of this increased resilience.

There are a number of factors which can affect the sizing of storage. First amongst these are disc access patterns and these must be allied to an understanding of the mix of transactions which the system has to handle. By looking at access patterns for various mixes of transactions, conclusions can be formed concerning the need to adjust the physical data mapping so that a storage strategy can be developed that minimises contention and yet maximises storage economy.

It is not possible to look just at the initial implementation and then leave the matter alone. Likely growth rates for the system need to be examined to show if problems are likely in the future. The disc drives may quite adequately cope with the message-rate now, but what will the situation be in one or two years' time? By getting early warning of any problems action can be taken to resolve or avoid them in good time.

This is a problem that never goes away! Once the system has been implemented and has stabilised, the predictions should be re-checked to ensure that the actual pattern of usage is not going to cause problems. The transaction mix will change over time and the situation should be periodically reviewed to check if any action is necessary.

With many systems certain key information is frequently accessed. So frequently, in many cases, that to keep this data on disc is bound to cause some performance overhead. A popular method for avoiding this problem is to lock this data into main-store memory so that access requests do not have to be queued while waiting for a physical read. Many of the high-performance database systems include facilities for achieving this and will manage the problems of refreshing from disc store as required. There is obviously a design trade-off here in terms of main memory availability and this needs to be assessed.

Redundancy

Various techniques can be used to allow for growth:

- *the database should be loosely 'packed' after initial load;*
- *allow spare room in record layouts where fixed-length records are a constraint;*
- *for network DBMSs define some 'dummy' sets to allow for additional relationships.*

Few database systems remain static. Over time it is almost universally true that new requirements will come to light and extra data types will be added. When designing the system some spare room to enable the system to grow should be allowed. Building in some controlled redundancy will help to lengthen the life of the system and make its maintenance and enhancement much easier to achieve.

When sizing the database this should allow additional space for growth in data volume. As a rule-of-thumb, the database should be no more than 50-60 per cent full after the initial load. This 'packing density' should only be exceeded when the data volume is known to be static and storage is at a premium. Once again, this free space will need to be periodically measured and as it starts to decrease, solutions to the inevitable filling-up of available space will need to be found. The initial sizing should allow space not just for data volume but also for index records, space management records where appropriate, and for physical pointers where these are used. Precise sizing rules and product-specific guidance should be sought from the vendor.

This exercise should also consider redundancy in record design. Many systems will only allow fixed-length records to be stored. Even in those that allow variable-length records the additional overhead incurred in handling these record types will sometimes preclude their use. If additional data items are needed on fixed-length records, there is an immediate problem as this is not possible unless the database is unloaded and reloaded and internal work space in programs, and so on, is changed to handle the extra fields.

This can be simplified. When the record layouts are initially designed it is useful to allow some spare bytes at the end of the record definition. This spare area can then be redefined, quite easily, to accommodate extra data items. By this means the record length remains constant and, apart from redefinition of the layout, all internal sizings remain unchanged. This technique can put off the day when there is no option but to begin unloading and reloading data. Be quite generous when allocating this extra space – it can be surprising how easily extra data is suddenly required.

Although applying only to Network DBMSs it is also necessary to allow space for the definition of additional relationships. In Network systems, space for pointers has to be allocated within each record type. If extra relationships need to be implemented, additional space will be required to hold the pointers to support that relationship. Defining redundant pointers is not as straightforward as defining additional data items. Many systems will not allow undefined pointers to be allocated. In this situation the problem can be circumvented by defining a number of relationships to a 'dummy' member record, none of which are needed in reality. By defining these additional relationships, the space taken by the pointers can be re-allocated when new relationships are required; once again, this becomes a matter of redefinition rather than unload/reload.

There are various other techniques, typically system dependent, that can be used to build in some redundancy. Consultation with the supplier of your database software will assist in making the appropriate allowances for your system. Some redundancy is always useful – but too much can become an overhead.

Reorganisation

Formulate outline plans for the following points:

- *which parts of the database are most likely to require reorganisation;*
- *what degree of change is anticipated?*
- *test out various scenarios on a test database of realistic size;*
- *consider ways to assist recovery from failures during the reorganisation process.*

However well planned the database is, and regardless of built-in redundancy, the day will come when it is necessary to reorganise or otherwise adjust the design of the database. Reorganisation is inevitable.

This can be a major problem for an up-and-running database system and it is therefore wise to consider the need for reorganisation from Day One. It is not unknown for a major reorganisation of a large established database to require several *days* of 24-hour-running to action. For many systems this can be an unacceptable amount of down-time! If the problem is recognised and anticipated early enough then plans to minimise the problem can be formulated.

The major factor in avoiding reorganisation is good initial design. This book has argued consistently for careful and thorough up-front design for the system. A good data model will be resilient to change and, by examining a wide scope at the start, extensions to the implemented system will be more easily accommodated. There is no substitute for this. Database systems cannot be grown "like Topsy" for, if the scope of the system is allowed to grow unchecked, the resulting problems will be difficult to overcome.

Redundancy can also make reorganisation less likely and less daunting. By building in adequate resource as discussed in the previous section, the problem will be reduced.

When all else fails, though, some form of reorganisation will have to be handled. Plan it now! Different DBMSs have different utilities available and the usability and benefits of these should be assessed. It is also essential to spend some time considering the type of reorganisation required. Will the existing storage structure need to be replanned and, if so, in what way?

Will the basic data structure need to be changed and how is migration to be managed with minimal disruption? Lastly, is the reorganisation merely to allow new record types and relationships to be 'bolted-on' to the existing structure or is the change more fundamental?

There are no right answers to these questions. They will depend upon the exact implementation and the degree of change involved, but, with consideration, it is usually possible to determine those areas most likely to need attention. When these have been isolated then definite plans can be made. It is necessary to assess how much data is involved, how many programs need amendment and the degree to which any vendor-supplied utilities can be used to assist the change.

Once such plans have been formulated, it is wise to test them out. Design a small scaled-down model of the production database and attempt to run the reorganisation against it. This will help to gather basic information on how long the reorganisation will take for the full-size system, how much work space is required to hold unloaded data and how easily programs and other procedures can be altered to run against the revised database. A final problem not to be ignored is coping with problems during the reorganisation itself. Can it be restarted in the event of a problem? If the run is going to take 18 hours and it fails after 17 (for whatever reason), can it be restarted without going back to square one? If the reorganisation is going to be run in a tight time-frame, and they usually are, this sort of data is vitally important in assessing the need for contingency and other emergency measures.

Housekeeping

The data archiving strategy must consider the following issues:

- *what volume of data requires archiving?*
- *how frequently must each type of data be archived?*
- *how should the archived data be held?*
- *should it be possible to restore archived data?*
- *how should restored data be re-archived?*

Housekeeping is an interesting area of database design that is frequently ignored by database designers! You should not fall into this trap since it is important to give full consideration to this area. One of the maxims of data analysis was that, in logical terms, there was an infinitely large database and that data, once created, existed for ever after. On the other hand, the process analysis did examine processes that cause deletion of data and the CRUD analysis went to some lengths to ensure that these were covered.

It is now time to act on this detail and to consider how long information should be retained and in what form. The form is important. Each data type will have two distinct lifespans, one overlapping the other. The first, and shorter, of these is the length of time the data needs to reside on disc and be available for immediate enquiry. This should have been established during data analysis and this, combined with volumetric details, allows the size of the database to be established. The second of these lifespans relates to how long the data needs to be retained in any form. The difference between the two defines how much data needs to be archived. Two situations can occur. The data lifespans may be the same: where this is the case, there is no need to archive the data type as it can simply be deleted once its expiry period is reached. Secondly, the lifespans will differ, in this case it is important to examine how the data is to be used once archived, and what volume of usage is envisaged.

This must be considered now. If a production system is implemented without paying attention to this area, the situation will arise where there is no storage left and there is no strategy for coping with the problem. Prior to this, the system will start to suffer significant performance degradation as the database fills up with out-of-date and largely irrelevant details.

How should this archiving be handled? The first thing to consider is whether details of what has been archived need to be recorded in the database itself. In a banking application, for example, there will probably be a great many account transactions which could be removed from on-line store fairly rapidly but which have to be held somewhere, perhaps for a considerable period, in case of query or complaint. In this situation it would also be useful to create some form of summary record giving details of those transactions archived when the details are removed. This summary record itself would provide some audit trail and a broad historical viewpoint that would be useful for enquiry.

So, it is possible to determine what data should be archived and when, and whether this needs to be recorded in the database. What about the archived data itself? How is this to be stored?

An initial view can be obtained by considering what has to be done if access to this information is required. Does it need to be restored so that it can be processed again, or is it simply needed for enquiry?

If only enquiry facilities are needed it may be possible to dump the data to microfiche or similar media which can be directly accessed where necessary. This is obviously very straightforward and should present little problem.

Sometimes, of course, data will need to be restored so that it can be subjected to some form of processing. Two factors need to be addressed here. It is not too difficult to write the data to magnetic tape in some reasonable form, but this has to be done in a way that enables the records to be read and updated back onto on-line storage. This means that all necessary keys must be available in the archived record to enable the retrieved data to be relocated in the database.

The sequence in which data is restored must also be considered. There is an immediate difficulty if account transactions are restored from a year-old archive, when the relevant account record was archived six months ago. Any plans have to ensure that data is restored in the right order. These problems have to be thought through and the appropriate strategy implemented. The problems are not insurmountable and the solution will depend on how much data needs to be archived, the complexity of the data structure and the fragmentation of the original archiving. Where different closely-related data types all have different on-line lifespans the problem is obviously more complex. In this situation there will be benefits in trying to rationalise the archiving strategy.

When data has to be brought back onto on-line storage, the next question is how to re-archive this data when, once again, this is no longer required. The archive strategy must not be compromised by writing the same data to several different archives, and the implications of this on any summary record also need to be thought through.

All in all, housekeeping questions are not always as straightforward or as simple as they may at first appear. The strategy needs to be considered at the outset as the need to design and implement good procedures can take considerable resource. Additionally, if summary records are required, these will affect the database structure. By addressing the questions given above, it should be possible to formulate and implement an effective strategy.

As with recovery strategies, the housekeeping procedures should be thoroughly tested before they are needed in the production systems. The strategy should be run against a test database and timings, volumes, and general usability assessed.

Chapter 20

Other Related Issues

User Staff

The more the users understand the system, the more tolerant they will be. Therefore:

- *ensure that user-staff do understand the basic workings of the system;*
- *try to appoint User Liaison Representatives and work closely with them;*
- *keep management informed of both problems and developments;*
- *remain responsive to user needs.*

It is now appropriate to move on from the purely technical issues of database design into some of the wider issues that are also important. This section (and to some extent the next) would apply equally to the implementation of traditional systems, but due to the fact that database systems can amount to a whole new philosophy, they have been included here as they require somewhat more emphasis. These two sections cover user-staff and systems-personnel.

The one golden rule with user-staff, if only there is one, is to get them adequately trained to operate the new system. Take this a little further. It is obviously important to cover basic operation of the system from their viewpoint and also to detail any manual or organisational changes that result, but the one thing that detracts from a successful implementation is fear of what that system actually *is*. People who are scared (as opposed to cautious) of technology generally do not understand it. This is not because they cannot understand it, but merely because the details have not been explained – so explain what a database is and how the system operates. This should not be a six-week course in computing but a basic understanding of the underlying nature of the system which is normally easily absorbed and can significantly 'de-mystify' the whole implementation. It is clearly beneficial if the users feel at home with the new system and are at their ease when using it. If they understand what they are doing they are more likely to trust the system, to use it correctly, and to get less upset when there are problems.

An excellent way of achieving and maintaining this communication with the users is by means of what can be called 'User Liaison Representatives'. These individuals, who would normally be first-line managers or supervisors, can help tremendously in achieving a good user-response.

Wherever possible, the developers should work very closely with these people. Their role is to have a thorough understanding of the system and how it should fit into the day-to-day operation of the business. Normally, these representatives have a line responsibility for ensuring successful implementation to their own management so it is useful to remember that they are on your side. Be honest with them. They can act as a very successful bridge between the development team and the end-users and they are a focal point in two directions. Firstly, they can provide valuable feed-back on how the system is being received and can report on any problems in an open-minded and unbiased way. Secondly, if they understand what the developers are trying to do and their problems, they can be a valuable mouthpiece back to the users. All systems have their problems and it is extremely beneficial to have an ally who is involved in the day-to-day operation of the system.

Whether there are suitable user representatives or not, it is imperative to keep users and their management fully informed of what is going on. Once again this must essentially be a two-way process: it must ensure that users understand the intentions and plans for the future; at the same time it should provide a vehicle for the users so that they can communicate their problems, requirements and priorities.

Systems and Programming Staff

The following issues require consideration for developers:

- *Training: how much and when?*
- *Standards: draft standards are essential, as is their review;*
- *'Deskilling': good project management will prevent this becoming a problem;*
- *Jealousy: Do not alienate other project teams.*

If implementing a database system represents a change in philosophy for the users, then this change is even more profound for the established data processing personnel. The new philosophy needs to be carefully explained because, unless it is fully understood, people cannot be expected to adapt to it in a positive and committed way.

The major means of countering this problem is by adequate training. The particular needs of individual project team members need to be determined and the correct level of training put in place. Wherever possible the training should be arranged just in advance of their joining the team. If training is arranged too early, then the lessons learnt will be forgotten by the time they need to use the skills. If the training is given after they have started to contribute to the project then, at best, they will find some of the training unhelpful; at worst, they will have already acquired bad habits which will be hard to correct. Many database vendors provide courses tailored to their individual products and these can be very useful for specific technical detail. For other issues, for example, data analysis, a product-independent course is probably more useful.

Another area which can help to ensure a good implementation is in developing an adequate set of standards to be used for project control. Standards should not be so cut-and-dried that they become restrictive, but a clear set of guidelines and an outline approach will provide a direction for people to follow. Just because the concept of database is new to the installation does not mean that it is impossible to formulate standards in advance. Database implementations share much common ground with traditional systems design even though some of the names and/or task responsibilities may have changed. Standards will need to be refined as more experience of practical issues is gained and this standards-review phase should be mandatory for each project.

Having a clear set of standards defined 'up-front' will prevent team members from making unilateral decisions and moving off in different directions.

However, there are also a number of personnel problems that are fairly common and need addressing. The first of these is the perceived 'deskilling' that is sometimes seen to be a result of a database implementation. Because a database implementation tends to force separation between the design of data storage and its manipulation, many systems analysts and even programmers can feel that the central administration of the database, its definition and control, removes an important aspect of their job. This is not really true. The analysts still have to state what data they need, if not the structure of it, and they are then free to concentrate on their primary and most interesting function, that of determining and designing the actual processes that have to be included in the system. For programmers, the situation is similar. They may feel that as they are no longer responsible for the file definition and its content, they are losing this valuable skill. Again, this is not really true. The programmers have been freed from these rather mundane tasks and can concentrate on producing the application itself.

Interestingly enough whilst the development team are feeling somewhat deskilled, other project teams may feel jealous of the same people. If the database project is seen as a major prestige development using new techniques, and it often is, then other staff members may feel left out. The project leader needs to be aware that these feelings can arise and be ready to deal with the resultant difficulties.

In general terms, this is really down to getting the approach across. Good public relations are as necessary within the data processing department as they are in the user area. There is nothing magical about the techniques being used, so there should be constant endeavour to de-mystify the whole situation and keep things in perspective!

Data Management and Database Administration

Data management and database administration are distinct roles:

- *Database Administration:*
  - *primarily technical;*
  - *day-to-day operation;*
  - *security, control and sizing etc.*
- *Data Management:*
  - *data 'Guardian';*
  - *organisation-wide responsibility;*
  - *independent of projects.*

How should the database be controlled and administered? This introduces the whole topic, not just of Database Administration, but also data management. Database Administration will be considered first.

This function is concerned with the day-to-day operation of the database itself. Aspects such as security and recovery of the database, physical sizing and the use of utilities are included in this function. This role is primarily technical and in many ways is analogous to the role of a systems programmer. The people who undertake this role need to understand the software and its operation; they must understand how the 'bits and bytes' operate and how to make the software perform to its best advantage.

What is data management? This is a more nebulous concept but basically embodies a common set of principles. In order to set up a central database, the data that resides on that database must be centrally managed. The structure and definition of the data is no longer the responsibility of individual project teams but belongs to the organisation as a whole. This is the natural extension of data analysis. Once management has decided to establish the database, a function needs to be established which is responsible for ensuring that the data model is converted into a working database and that any difficulties over the definition of this data are resolved. This function embodies the 'Guardian' of the organisation's data and this must be established as an on-going and responsible position.

This function is of critical importance to an on-going successful implementation and therefore the skills needed are essential. The person or persons who would staff this role must have a thorough understanding of the business sector, the organisation and its data. Tact and diplomacy are indispensable.

This role needs to moderate and resolve any difficulties over what the data means and how it should be used. Technical skills are less important if the Database Administration function is technically strong. However, there is a need for database techniques and their application to be understood.

The positioning of these two roles within the organisation structure is also a key factory. All-too-often they are buried with the systems programming or technical support functions. This simply will not work! The data management function has global *responsibility* for company data and must also have sufficient *authority* to carry out its role and implement its decisions.

In many organisations this can be successfully implemented within the systems function but, if it is, it should be responsible directly to the highest level of systems management. The function must not get embroiled in normal project-based structures. If this happens, the function can get bogged-down in project-oriented pressures and timescales. The function must be separated from these problems and needs room to work. The other option is to establish the function in parallel with the systems department. This can work well but the close cooperation required between the systems department and data management can be more difficult to maintain. Generally, the most effective structure is to place data management within the systems department but independently of technical support, operations and project development.

Even with a well-constituted data management function it is still necessary to establish an umbrella of control for setting priorities and providing direction. To reflect the business-oriented role of this function this can be well catered for by setting up a steering committee, at a senior management level, to resolve difficulties and provide an organisation-wide view of what needs to be done and when.

Such a committee would have a senior manager from each of the user-areas that make use of the central database. The data management function would also be represented and, most importantly, there should be a chairman, at a higher level of authority, who would be seen as being independent and able to command the respect of the other committee members. This type of structure allows data management to have a clear reporting line back to the users and to have sufficient authority to achieve its primary role of data 'Guardian'.

The terms of reference, both of the steering committee and data management, should be carefully specified and agreed. As data management is often a new and challenging concept, if it is not given clear guidelines in which to operate it is likely to fail to deliver. Data management can be easily hamstrung by users or project teams that do not wish to take a corporate view. By placing data management at the right level within the company it can be given the authority and independence it needs to do the job!

Data Dictionaries

Data dictionaries allow the storing of 'data about data' (meta-data). Typically, they contain four major areas:

- *Business Data: the results of data analysis;*
- *Business Processes: the output from process analysis;*
- *Computer Data: details of files, records and data fields etc;*
- *Computer Processes: data concerning systems, programs and other routines.*

Purchasing database software is rarely the end of the story. For many installations there is a vast array of related software that is often purchased to complement the base product. The remaining three sections of this chapter describe the main types of related software and give some information as to their use. First amongst these are Data Dictionaires.

So what is a Data Dictionary and how is it used? Traditionally, data processing departments have been paper intensive: all the specifications, file layouts and source code have been produced and stored on paper. Although paper has some advantages in use, it cannot be readily accessed by machine and therefore the technology that has been delivered to users *via* computerisation has not been applied to DP.

This rather hypocritical state of affairs is beginning to change. Data dictionary software provides the data processing department with a means of storing and accessing data about data – or meta-data. Data dictionaries are typically the result of carrying out data analysis where the user is the data processing department, and then converting this into a database application. The power and flexibility such systems require has meant that database techniques have been needed to actually build the application, although there is no need to have a database application in order to benefit from the advantages they bring.

So how is a data dictionary used and what meta-data can be stored? The range and capability of data dictionaries varies from one product to another, but in general terms they allow the recording of information in up to four distinct areas. These can be referred to as the four quadrants of a data dictionary and are shown diagrammatically in Figure 20.1.

As can be seen, the diagram is divided horizontally into two areas, the Business World and the Computer World. Additionally there is a vertical split between Processes and Data. This type of structure allows data to be recorded about most of the things relevant to system design. Figure 20.2. takes this further and shows typical entity types that may be held in each quadrant. It can be seen that the results of Process Analysis belong in the Business Processes quadrant, whilst the data analysis results are held in the

BUSINESS PROCESSES	BUSINESS DATA
COMPUTER PROCESSES	COMPUTER DATA

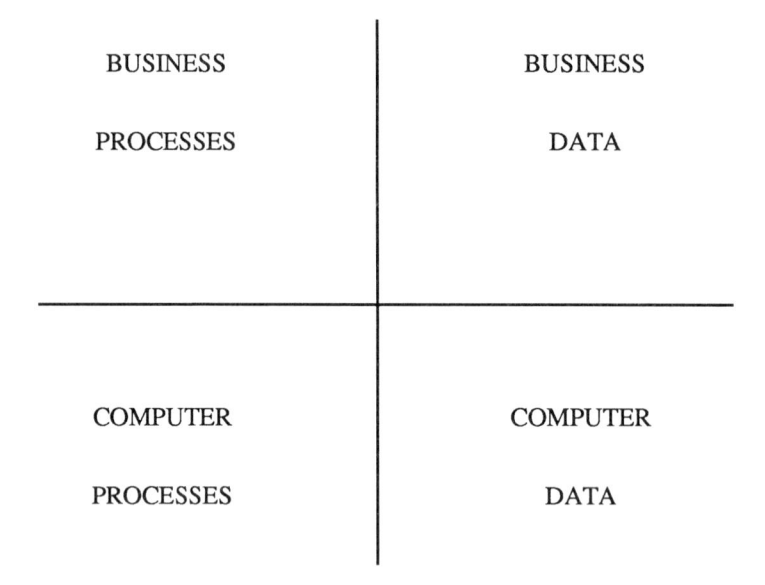

Figure 20.1   The four Quadrants of a Data Dictionary

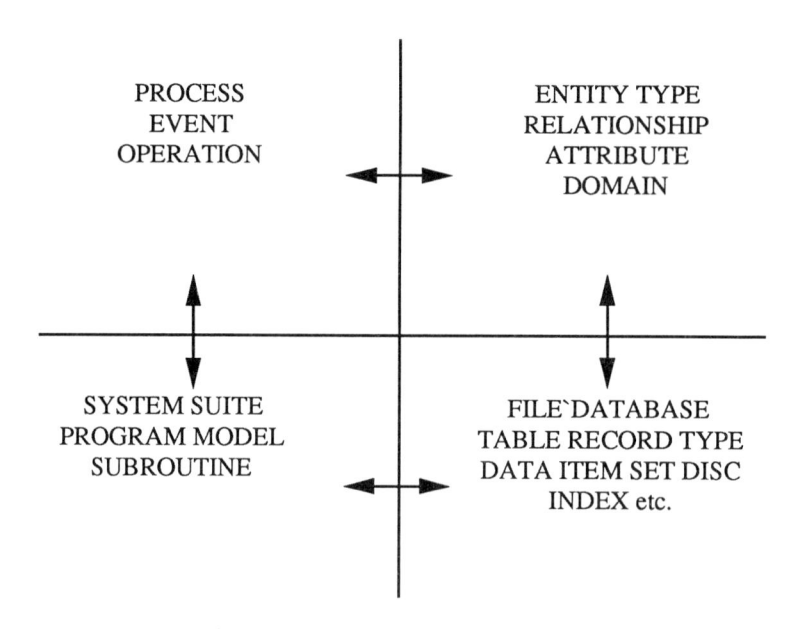

PROCESS EVENT OPERATION	ENTITY TYPE RELATIONSHIP ATTRIBUTE DOMAIN
SYSTEM SUITE PROGRAM MODEL SUBROUTINE	FILE`DATABASE TABLE RECORD TYPE DATA ITEM SET DISC INDEX etc.

Figure 20.2   Typical Data Dictionary Entity Types by Quadrant

Business Data quadrant. A sophisticated data dictionary will also allow cross-referencing between the two. At the lower level is the Computer world and this can record details of systems and programs in Computer Processes and the implementation of the data model in Computer Data. Again, some products allow these two quadrants to be cross-referenced. The mapping between the Business and Computer worlds is very important and it is, therefore, useful to have effective cross-referencing capabilities between the two Process quadrants and the Data quadrants.

To give a flavour for the structure of such software (often a database implementation itself), a simplified data model for the Business Data quadrant is shown in Figure 20.3. A full implementation would need a good deal more complexity but this may help to put the dictionary concept into perspective.

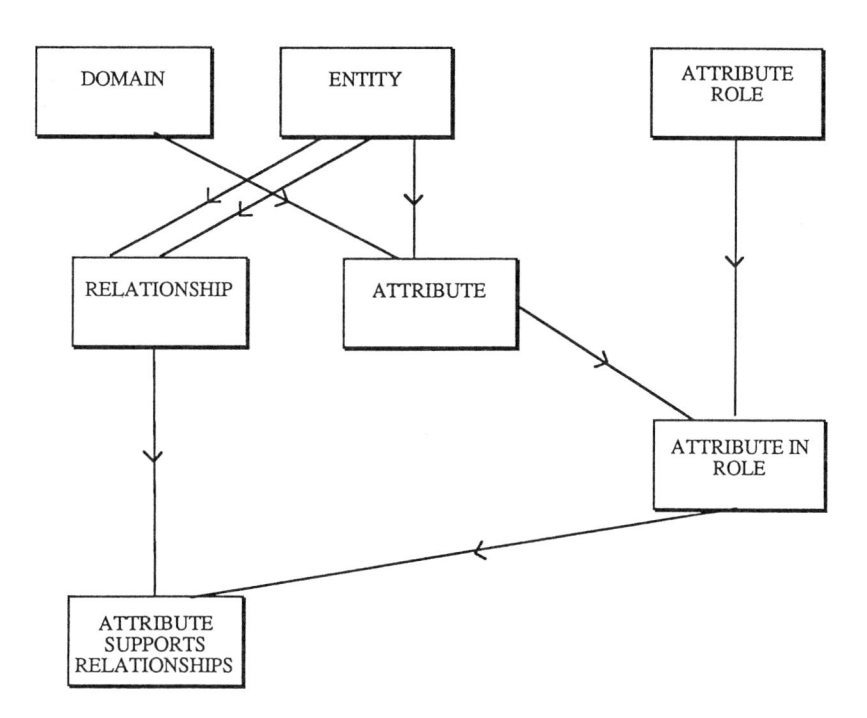

Figure 20.3 Simple Data Model for the Business Data Quadrant

The advantages of using a data dictionary are numerous. The major advantage is the ability to store all this meta-data in one place and to maintain consistent and clear definitions of data and its use, globally, for the whole department. Then by providing flexible and powerful enquiry facilities, the Dictionary allows the user to discover what a piece of data means, where it is used, and the implications of changing it. Thus, if it is necessary to change a particular record layout, it should be possible to trace through and discover which programs use it, what files or databases the record is on, and the effects that this change may have.

This can usually be achieved with a few simple but powerful enquiries and saving much of the searching through program and system specifications that would be necessary for traditional methods of documentation.

The primary disadvantage of data dictionaires is persuading people to use them and keep them up-to-date. This is a problem with any form of documentation, but data processing professionals seem to be remarkably resistant to taking their own medicine!

This problem can be eased where the dictionary is sufficiently powerful to yield other benefits. Because the definition of data is held within the dictionary, some products allow these definitions to be used directly by programs. The programmer does not need to include data definitions explicitly but simply calls these directly from the dictionary. By such means, manual alteration is no longer required when things change as the revised layout can be obtained automatically.

Some vendors take this even further by insisting that the dictionary is used to control the system. For example, a program that is not known to the dictionary may not be allowed to run or access file definitions to which it does not have authorisation. Additionally, some systems will generate first-cut database designs from the business level information and by this means they become an integral and necessary part of the software environment.

Finally, it is becoming increasingly common for the data dictionary actually to support and drive a design methodology. As the project develops the results of data and process analysis are fed into the dictionary which is then able to look for inconsistencies and other problems, prompt for missing information, and generally assist in ensuring that a full and complete analysis is carried out.

When the analysis is complete the dictionary can generate a rough implementation which can be customised and tailored to suit specific requirements.

These developments are very exciting and introduce a much more professional approach to systems design. It is almost certainly the way of the future – so come to terms with it now!

Query Languages

Using a query language is not an answer to all problems especially:

- *they can be inefficient;*
- *syntax may be awkward for casual users;*
- *careful control of availability required;*
- *they are at their most useful when closely integrated with the database and data dictionary.*

Query language is another area of software development that has grown in popularity and is often closely associated with database techniques. The basic function of these facilities is to provide a means to make *ad-hoc* enquiries about data in a simple and non-technical way. The languages are frequently non-procedural, that is, they specify *what* data to retrieve but not *how* to retrieve it. As such they achieve a measure of independence between the process and the data. The physical structure and location of the data may be altered but, providing the logical content of the data is the same, the query should yield the same result. The most well-known of these languages, and now an accepted standard, is Structured Query Language (SQL) from IBM.

As the language does not specify where the data is, nor its structure, these query languages are often heavily interrelated with data dictionaries: it is the dictionary that 'knows' where the data is and how it is structured. This is another area that is making the use of dictionaries obligatory and adds weight to the argument for making the dictionary the central piece of software to control and administer systems.

The major advantage of query languages is that they allow non-computer professionals access to the data. A program, as such, need not be written. Provided the user knows *what* data is held within the system and is conversant with the syntax of the language, then *ad-hoc* enquiries can be made quickly and easily with little or no pre-planning. By this means, many simple enquiry and reporting requirements can be satisfied by the user directly, thus saving valuable resources for the more complex and demanding applications.

However, query languages are not without their problems. First of all, they can gobble-up tremendous machine resources.

The user is unaware of how complex the solution to the query may be or of how efficiently it will run. Also, the temptation is to 'browse' through the data, possibly repeating the query several times in slightly different ways in the light of each response. As a response to this some software products make use of query optimisers to reduce inefficiency. The optimiser typically examines the query and by reading the dictionary (what else?) assesses several possible strategies for satisfying the query in order to determine the most efficient. This helps but is not an answer in itself. A further refinement is to return to the user some form of cost estimate or execution plan prior to actioning the request. This gives the user a chance to abort the query if it is too expensive or will run too slowly for their needs. With many languages, a slight re-adjustment of the query arguments can lead to massive differences in efficiency. In the hands of an experienced user, feedback on the cost will enable the user to tune the enquiry prior to execution.

The other major problem lies in telling the user what data is available and what it is called. Again, some products, using close cooperation between the dictionary and the query facility, will prompt or help the user by providing details of what data they may look at and how that data is defined. This is also of assistance with the syntax of the language; SQL and its equivalents, are not normally the most natural forms of English and for an occasional or disinterested user would probably be totally demotivating. This can be overcome with facilities that allow the user to complete query menus and use pop-up windows, and so on, to prompt and detail the options available.

The final major area concerned with query languages is their control and administration. Careful thought needs to be given as to how such powerful and unstructured tools are to be made available to end users. It is also necessary to set up thorough privacy constraints on what can be seen.

The availability of the query facilities should also be controlled. As already stated, these facilities can literally run away with machine resources. If the facility is over-used, response will be poor and front-line application systems may be affected. The facility needs to be monitored and restricted if this is necessary to maintain the quality of service. Where possible, cross-charge users for their use of the facility: this will promote a greater sense of responsibility and will certainly prevent a user doing continuous serial reads of the entire database just for fun!

Lastly, user education is very important - they have to know how to use the facility and where to ask for help and advice. Make sure a suitable line of assistance is available and adequately staffed. The user also has to understand the data if the results of the query are to make sense. Users need to be kept informed of what data is available and what it means. Query languages are very useful but they are not a panacea!

Other Data Management Software

Additional tools which may be available include:

- *Report Generators: simplifying report production;*
- *4GL environments: high productivity languages for process code generation;*
- *Screen Builders;*
- *Application Generators;*
- *Database 'Fixers' for correcting corrupt data.*

Many of these tools are closely interrelated with the underlying data dictionary.

Allied to data dictionaries and query languages are a number of other software products that, when used in conjunction with one another, make up a complete software environment.

The first of these, and in some ways closely allied to the query language, are Report Generators. Whereas a query language is primarily aimed at providing a screen-based interactive response, a report generator is aimed at providing similar facilities in a printed form. Producing and maintaining programs to generate reports takes up an awful lot of designer and programmer time, yet this type of program should be relatively easy to produce. The rules can normally be defined in a stylised way similar to query languages and only extra features such as control breaks, headings and totalling facilities need to be added. In many systems these reports can either be delivered to screen or printer allowing the user greater choice in utilising the information. Most reporting requirements can be satisfied by report generators and they typically yield a much higher rate of productivity over traditional languages. Once again report generators can be made available to users directly, but the same administrative issues will then need to be addressed.

Apart from removing certain mundane tasks, query languages and report generators do not help the programmer very much, but there are a number of development aids available. These are normally wrapped up under the banner title of fourth generation languages (4-GL). These languages are typically non-procedural and enforce some degree of program-data independence, but include data manipulation facilities such as would be found in Cobol or Assembler. High productivity gains are often claimed for these languages, but it is necessary to provide thorough training to effect the change in approach that is required.

These languages allow the programmer to produce the basic modules of processing code that are essential in any sophisticated system. Enquiring on data is one thing, manipulating it and putting it to use is something else.

Given that the basic building blocks can be produced with the 4-GL we have two further aspects to improve.

The first of these are screen-building techniques. Some of these are now getting very sophisticated, allowing default screens to be produced instantaneously given the format of an SQL-like statement. Having produced the default this can often be customised by the developer to add specific headings or text, to rearrange fields on the screen, and generally to improve the usability and presentation of the screen for the end-user.

The final area to address is actually building and structuring the application. There are tools to help to define the screens, generate the database, set up query and reporting facilities, and create process-code building blocks. All that is needed now is to string them all together into an application. Once again, modern software facilities exist to help. Application Generators allow processes to be attached to screens, screens to be sequenced, and menus and choices governing the route through the application to be generated.

All these facilities tend to make extensive use of the data dictionary and may update the dictionary so that this contains complete documentation on the structure of the application. By separating out the various parts of the application, these software facilities allow great flexibility and ease of change. This gives the systems department the opportunity to provide the right systems quickly and easily.

There is one final area that we have not addressed. Sooner or later there will be a corruption of the database. The Database Administrator needs a global facility to get at and change corrupt data without the constraints imposed by a formal application. These database fixers are available for some products, although not all. They are invaluable, especially for large databases, but they are easily abused. It is necessary to put some control around these facilities by logging their use and ensuring that the facility is well protected from casual use. The ability to cut through all privacy constraints and alter data, often without leaving an audit trail, is a powerful but dangerous thing. Do make sure that controls are water-tight, preferably restricting the use of such facilities to one or two key individuals.

Chapter 21

The Way Forward

Why Use Data Analysis?

Data analysis confers the following benefits:
- *data-driven to lead to more flexible systems;*
- *it is a mature, proven approach;*
- *largely self-documenting;*
- *ensures user involvement through iterative design;*
- *business not systems oriented.*

This book has covered a lot of ground in the foregoing sections. The techniques and use of data analysis have been examined and how these techniques can be used to design better, more flexible systems has been discussed. In later stages it discussed how the results of this design can be converted into an implemented solution and examined the major issues that must to be considered when doing this. In this final chapter then, it remains to try and pull together some of the threads that run throughout the foregoing discussions and look to the future to see where this philosophy is likely to lead. This first section reviews why data analysis should be used and re-emphasises some of its fundamental benefits. The next section looks at Database techniques and their application. The penultimate section looks to future likely trends and developments; and, to tie the whole thing together, the final section contains some concluding remarks.

So, as a reminder of the overall objective, why use data analysis? Data analysis has definite aims and clearly defined techniques. These, more than anything else, allow a clear, logical and thorough examination of the business area to be made. Although there is an underlying philiosophical change to be made by the analyst, once this view of establishing the data structure has been accommodated, the use of data analysis is not particularly difficult. This change in philosophy is important. The traditional process-oriented view of systems analysis has led to the development of systems that are highly process specific and rigid. By adopting the more generalised data-driven approach, it is possible to reach a level of understanding that is more fundamental and this allows the production of systems that are more flexible and responsive than was previously possible.

Data analysis is not new. Although understanding of the power of the technique is still expanding, it is mature enough to have been shown to work and to have become a trusted and generally accepted technique. The techniques are now part of most accepted methodologies and are known to give benefits; the difficulties of actually implementing systems that can reflect the power of data analysis are also rapidly being overcome.

The techniques are largely self-documenting in themselves. By describing and defining entity types and processes, the major content of traditional user requirements and systems specifications is covered in a thorough and consistent way. The standards laid down in this text allow and encourage a complete specification of the business to be built up and understood.

Perhaps one of the most important features of data analysis is the degree of user involvement required. Yes, it is possible to sit in an ivory tower and produce a data model, but it will not be very useful. To make the exercise work, consistent user involvement is essential and the iterative nature of the design process encourages this involvement. This may, of course, come as a shock to the users. But if you want an individually designed house, you must spend much time with the architect! If you do not, go out and buy a package and live with someone else's view of your business.

Lastly, data analysis is business rather than systems-oriented. This ensures that a thorough understanding of the business is obtained before embarking on systems design with all the trade-offs and compromises that that entails. At last computer application design is starting to mature. It is now possible to check back to basics as the task is progressed. Data analysis provides an essential first step in this process and one that has been poorly served by traditional methods.

DBMS or not DBMS?

Database techniques confer their own benefits:
- *easier development of flexible systems;*
- *single sourcing of data;*
- *controlled definition and use of data;*
- *enables information provision as an objective in its own right;*
- *makes data more accessible.*

For many installations, this indeed is the question! Although this book is not concerned with databases *per se*, it is essential to draw the parallels between data analysis at the business level, and database solutions at the computer level.

Databases do give many advantages and, as the technology improves, these are becoming more wide-reaching in their applicability. As a means of implementing a data model without major compromise, they are indispensable and enable the flexibility and elegance of the data model to be carried through to the implementation level. The main thrust of this, of course, is the concept of data being held and maintained for the benefit of all users. The independence of data is enabled by developing and implementing a database giving what might be termed 'one-source' controllability. The data manager knows and understands the data needed by the organisation. The data is consistently defined and made available to the appropriate user in the appropriate way. In essence, the application does not define the data, the application only defines a 'view' and this, of course, may change. It is rare for the underlying data to change. The integration of the data provides many benefits to the organisation rendering it possible to rely on the quality and meaning of data without having to understand the nature of the source system.

Many installations are moving towards the establishment of Information Centres or similar functions. This is a response to the changing nature of computing: more and more frequently, user requests are moving out of the day-to-day 'Data Processing' arena into the more flexible and unpredictable area of 'Information Provision'. The day-to-day processing still has to be covered but management needs information and free access to it. To use information properly it must be consistently defined and understood, and the appropriate software tools need to be in place to permit this access. The use of query languages and similar facilities is beginning to set users 'free'. Data is being returned to its rightful place, in the hands of the users themselves.

Future Scenarios

It would seem probable that two different streams of data processing will emerge:

- *'Traditional' process-oriented development: this will address the major operational needs of the organisation;*
- *Information Provision: specialists in the use of the software to provide flexible and powerful development tools for end-user computing.*

Where does all this lead? It would seem that there is increasing polarisation within the systems department. On the one hand there will be the continuing need to build traditional production systems to drive the day-to-day affairs of the organisation. There will be changes in the software, large-scale efficient database software will become more and more widespread and the use of Application Generators will become commonplace. In basic terms, this will still be the home of the traditional systems analyst and programmer. The building of pre-designed fairly rigid applications will continue.

This type of system, though, will not meet the ever-increasing needs of users for reliable and flexible information delivery. Here the information centre concept is beginning to come into its own. The information centre is there to provide the user with the tools and support that is needed. Flexible and powerful enquiry and access facilities will need to be made available, not by traditional applications staff, but by technicians who are skilled in the software and understand *how* it can be used rather than *what* it can be used for. The user will gain more and more expertise and take more responsibility for implementing solutions. The technicians, a specialised form of systems programmer, will be there to advise and guide the user and will get involved only with the production of the more complex requirements.

Whether these facilities run on the main database or on a specialised enquiry database is a technical rather than business decision. The concepts remain unchanged. The 'pot' of structured corporate data needs to be controlled and administered by computer professionals, but the *use* of that 'pot' is limited only by the imagination and enthusiasm of the user. This is inherently right. The systems department is there to provide the means and the facilities needed by the user: it is not their role to dictate what can be done.

As software aids grow in power and usability and as users become more computer-literate, these trends are bound to continue. This is not a threat to the computer department, but it does represent a challenge to which we must all rise! Techniques such as data analysis will be even more important as this change evolves. The need to determine the data and its structure will remain; indeed the usability of the data will depend entirely on how well this is done. Let the users define the processes, they know what they want to do, but get the data organised and under control for the benefit of all.

In Conclusion

There is an ever-burgeoning need for information. These techniques enable data to be used effectively and turned into timely, accurate and reliable information.

The foregoing text has presented an integrated set of techniques and software. We now have the technology, and with it have come the techniques and approaches needed to design and implement powerful information systems designed to respond to users' needs.

This is fundamental. These techniques can be used to address real problems and provide practical solutions in a manner hitherto impossible. It is fair to say that this is only the beginning of the road – but at least it does seem to be the right road! Where this will lead in the future is unknown, but there seems to be no end to the information requirements of users. The more information supplied, the more is needed.

It is now possible to give back to users the freedom to access information that they lost when data was first computerised. The freedom to examine, enquire about and use this information will itself give rise to the need for more systems and even more powerful software. Expert systems will grow to meet the need to process vast quantities of data quickly and flexibly.

Time and time again this comes back to the basic need to understand and control this data. It is an essential pre-requisite without which none of these developments will be possible.

Data analysis is the key. It can be used to unlock an array of challenging and exciting possibilities. By rising to meet this challenge, the power of the technology will benefit us all.

Appendix

Sample documentation

INTRODUCTION

This Appendix includes some sample documentation which illustrates the typical output of a data/process analysis exercise. The appendix is in two parts. Data analysis outputs are given first followed by details of process analysis outputs and the mapping of the processes to the data.

Certain points should be noted:

- The documents are illustrative. The examples are fictional but are based on a real-life example, although somewhat simplified;
- The examples are samples only. A full set of outputs is not provided as this would be too bulky and somewhat tedious. However, the samples should be consistent enough to allow the reader to follow it through;
- The examples are drawn from a working model, that is. the analysis is not complete. This is valuable as it shows the way in which uncertainties and unknowns can be documented for later resolution;
- As the model is not finished there are areas of uncertainty and also areas that could be improved. The reader may care to try to identify these and resolve these uncertainties.

DATA ANALYSIS DOCUMENTATION

Data analysis outputs consist of three parts:
- The data model diagram;
- A textual introduction to the main concepts;
- Sample entity type descriptions.

DATA MODEL : Motor Insurance Application

Author : Patric Downes

Date : January 12 1989

MAIN CONCEPTS

1 Vehicles

Each manufacturer is represented as an occurrence of MAKE with each specific model set up in the MODEL entity type. The MODEL occurrences are categorised by TYPE OF VEHICLE to distinguish between private or Commercial etc. Specific vehicles are represented in the VEHICLE entity type and related back to MODEL. Any special factors related to the vehicle are held in VEHICLE MODIFICATION.

2 Policies

The basic policy details are held in POLICY and these are related to the appropriate INSURER. POLICIES will always have a related POLICY STATUS which can change over time. The history of these changes is held in STATUS HISTORY. The SPECIAL INFORMATION entity type is included to contain details of any notes or instructions peculiar to that POLICY. There are two reference tables related to each INSURER. These are AREA CODE which defines the Insurers risk areas and also VEHICLE CODE which defines the range of insurance groups relevant to the INSURER. Specific MODELS are related to the VEHICLE CODE via MODEL CODE.

3 Cover

The variable terms of insurance for each POLICY are contained in the COVER entity type. This has a related COVER TYPE which primarily specifies the overall level of cover available. The VEHICLE or VEHICLES insured under the POLICY are related to the COVER via VEHICLE ON COVER. All known drivers, including the policyholder, are defined in the DRIVER entity type. Any special driver history, for example, accidents, convictions or disabilities is detailed in DRIVER EVENT. These are classified by type, accident, conviction, disability, by EVENT CODE.

4 Customers

Any party related to a POLICY is held in the CUSTOMER entity type. The address, which can obviously change, is held in ADDRESS. Each ADDRESS is related back to risk areas via ADDRESS IN AREA. All CUSTOMERS are related back to the POLICY with which they are involved via INTEREST IN POLICY. This entity type is categorised by INTEREST TYPE to identify the manner of the CUSTOMERS involvement.

5 Payments

Sums due on a POLICY are held in PAYMENT DUE. Any payments received from a CUSTOMER are held in PAYMENT RECEIVED. The cross-reference CASH ALLOCATION allows payments and debits to be cross-related as necessary.

6 Other areas

A number of areas still require exploration and cannot yet be modelled.

These include:

* Claims and their history;

* Correspondence including Cover Notes, letters etc.;

* Diarising of follow-up actions.

DATA MODEL : Motor Insurance Application   Page 2

Author : Patric Downes

Date : January 12 1989

PARTIAL DATA MODEL

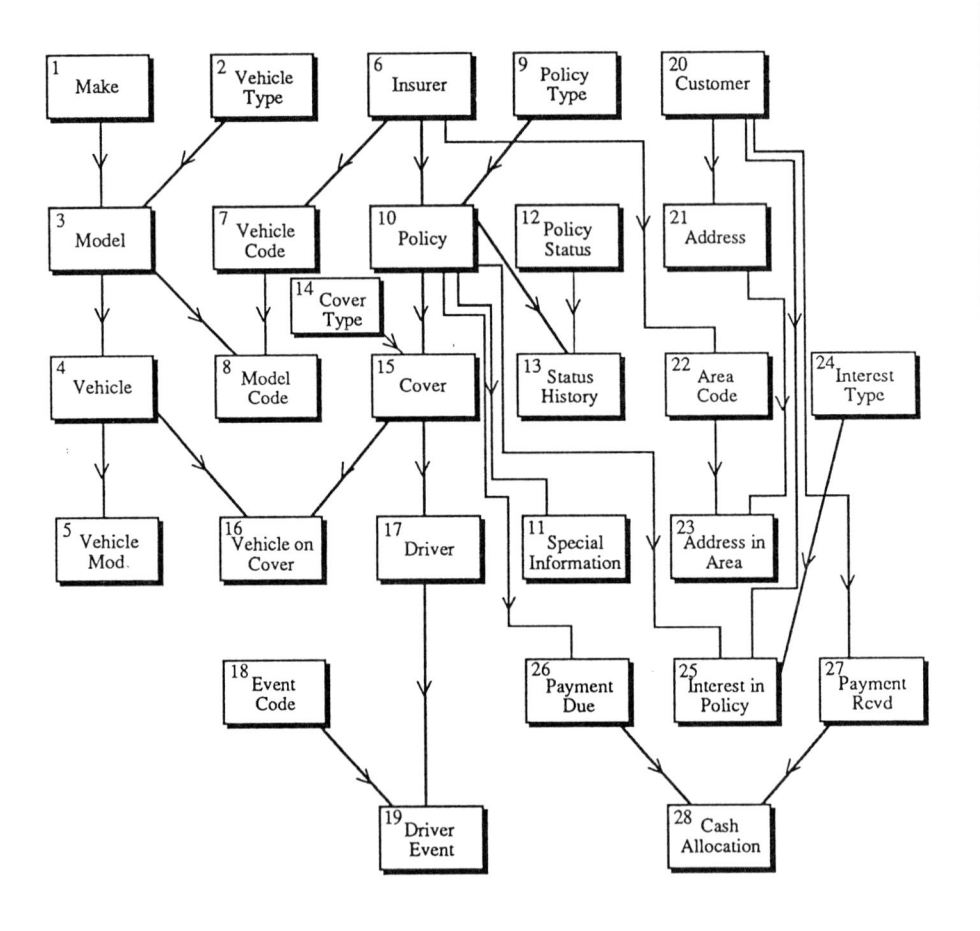

1. Make
2. Vehicle Type
3. Model
4. Vehicle
5. Vehicle Mod.
6. Insurer
7. Vehicle Code
8. Model Code
9. Policy Type
10. Policy
11. Special Information
12. Policy Status
13. Status History
14. Cover Type
15. Cover
16. Vehicle on Cover
17. Driver
18. Event Code
19. Driver Event
20. Customer
21. Address
22. Area Code
23. Address in Area
24. Interest Type
25. Interest in Policy
26. Payment Due
27. Payment Rcvd
28. Cash Allocation

Practical Data Analysis

DATA MODEL : Motor Insurance Application

Page 8

Author : Patric Downes

Date : 14 January 1989

ENTITY TYPE: 6 INSURER

Description

This entity type defines each INSURER with whom the company places POLICIES. An INSURER may be either a limited company or a syndicate of underwriters at Lloyds.

Attributes

INSURER CODE	INSURER NAME	ADDRESS	TEL. NO	MAIN CONTACT
I				
UK	UK Motor Policies	UK House, Bishopsgate London EC3	01-623 9167	John Smith
FJ	Fred Jarvis Lloyds Policies	16 Station Rd Stratford, London E15	01-472 1655	Bill Bloggs
CV	Coverall Ltd	77 Great Scott St Glasgow	031-215 9696	Hugh McNish

Illustrative, Hypothetical

Volumetrics

Initial:	Currently 10 underwriters are used.
Growth:	Unlikely to exceed 1 per year.
Retention:	Retain until 6 years after expiry of last expiring policy.
Relationship Population:	N/A.

DATA MODEL : Motor Insurance Application

Author : Patric Downes

Date : 14 January 1989

ENTITY TYPE: 6   INSURER, continued

Source

Initial: Existing records (Insurance Admin)

Ongoing: Head of group (1 per insurer).

Questions

1   Would it be useful to distinguish between different INSURER types?

2   Is one contact sufficient, or are there contacts for specific purposes?

DATA MODEL : Motor Insurance Application   Page 12

Author : Patric Downes

Date : 31 January 1989

ENTITY TYPE: 10 - POLICY

Desription

The POLICY entity type contains the 'fixed' data concerning each issued POLICY. It represents the legal undertaking of risk by the INSURER by whom it is issued. Each POLICY must be of a specific POLICY TYPE.

Attributes

POLICY NO.	INS. CODE	POLICY TYPE	START DATE	RENEWAL CYCLE	EXPIRY DATE
I	S	S			
MP10116	UK	PRIV	01-01-86	ANN	-
P7166A	CV	PRIV	25-10-88	ANN	-
P1001B	UK	COMM	19-03-84	6M	18-03-86
AX75BJ	FJ	PRIV	21-10-85	ANN	-

Illustrative, Hypothetical

Volumetrics

Initial:	Currently about 100,000 in force, with 200,000 lapsed.
Growth:	About 100/week = 5,000/annum.
Retention:	Retain for 6 years after 'Expired Date'
Relationship Population:	To INSURER:
	One major INSURER covers approx 50 per cent of all POLICIES.
	The remaining insurers have about equal coverage.

Sample Documentation

The remaining insurers have about equal coverage.

DATA MODEL : Motor Insurance Application Page 13

Author :   Patric Downes

Date :   31 January 1989

ENTITY TYPE: 10 - POLICY, continued

Relationship population, continued

> To POLICY TYPE:
>
> 80 per cent of POLICIES are 'Private' (PRIV).
> 15 per cent are commercial (COMM) and the remainder are motor trade (TRAD).

Source

Initial: Admin records.

Ongoing: New Business Clerks.

Questions

1. Is the 'Renewal Cycle' useful, or are all POLICIES issued for 12 months?

2. Will the 'Expiry Date' always equal the 'End Date' of the most recent COVER entity or is it possible for the POLICY to remain in force when cover is suspended?

DATA MODEL : Motor Insurance Application Page 16

Author : Patric Downes

Date : 3 February 1989

ENTITY TYPE: 15 - COVER

Description

The Terms and Conditions defining each instance of risk are held in the COVER entity type. Each time the risk on a POLICY changes, a new occurrence of COVER will be created. This will include any policy adjustment as well as periodic renewal. COVER occurrence must be of a specified COVER TYPE indicating the general level of cover provided, eg. Fully Comprehensive, Third Party only etc.

Attributes

POLICY NO	COVER START	COVER END	COVER TYPE	NO-CLAIMS YEARS	RESTRICTED DRIVERS	PROT NCD?	USAGE	ANN PREM(£)
I/S	I		S					
MP10116	01-01-86	31-12-86	COMP	3	P	N	B1	127.40
MP10116	01-01-87	09-05-87	COMP	4	P	N	B1	140.50
MP10116	10-05-87	31-12-87	TPFT	4	P+S	N	B2	98.75
AX75BJ	21-10-85	20-10-86	TPO	0	A	N	SD	70.27
AX75BJ	21-10-86	20-10-87	TPO	1	A	N	SDP	68.00

Illustrative, Hypothetical

Volumetrics

Initial: 6 years history is required. Over the past six years, 18,000 POLICIES have been incepted. The average POLICY life is three years (requiring 3 COVERS) with approx 40 per cent of POLICIES having an adjustment each year/mid-term. The approx. volume is therefore 18,000 x 3 x 140% = <u>75,600</u>.

Sample Documentation

DATA MODEL : Motor Insurance Application

Author : Patric Downes

Date : 3 February 1989

ENTITY TYPE: 15 - COVER, continued

Growth:	At current growth rates:	One per new policy =	5,000
		One per renewal =	10,000
			15,000
		Less lapsed per year	3,500
			1,500
		Adjustments (40%)	4,600
		TOTAL	16,100
			per annum

Retention: COVER details should be retained for six years after 'End Date'.

Relationship Population:

    To POLICY: Each POLICY will have an average 4 COVERS.

    To COVER TYPE Approx 25% are TPFT

          60% are COMP

          15% are TPO.

Source

Initial: Current: from Insurance Admin.

        History: from archives.

Ongoing: Depends on reason. Includes New Business, Adjustments and Renewals sections.

Questions

1. Are there specific code values for 'Usage' and 'Restricted Drivers'?

2. Does the 'Usage' attribute refer to the COVER, or is it specific to each insured VEHICLE ie. is it possible to have two VEHICLES with different terms on one POLICY?

3. Is it feasible that the COVER on a POLICY would be changed more than once in a day?

DATA MODEL : Motor Insurance Application   Page 19

Author : Patric Downes

Date : 3 February 1989

ENTITY TYPE : 17 - DRIVER

Description

The DRIVER entity type contains details relating to each known DRIVER under a POLICY. The DRIVER will include Policyholder details where the holder is also a DRIVER. Only one main DRIVER is permitted at any time on any POLICY. The indicated main DRIVER is always used for risk assessment purposes. Where the COVER permits 'Any DRIVER', the DRIVER entity occurrences will detail all declared DRIVERS.

Attributes

POLICY NO. I/S	COVER START I/S	DRIVER NO. I	DRIVER NAME	DATE OF BIRTH	DRIVING SINCE	RESIDENT SINCE	SEX	MAIN DRIVER
MP10116	01-01-86	1	Policyholder	19-04-53	1974	1953	M	Y
MP10116	01-01-87	1	Policyholder	19-04-53	1974	1953	M	Y
MP10116	10-05-87	1	Policyholder	19-04-53	1974	1953	M	N
MP10116	10-05-87	2	Mary Kelly	27-10-54	1978	1954	F	Y

Illustrative, Hypothetical

Volumetrics

Initial: On average there are 1.5 DRIVERS per COVER. The initial volumes for full history is 75,600 + 50% = <u>113,400</u>

Growth: Based on growth in cover occurrences: 16,100 + 50% = <u>24,150</u> per annum.

Retention: DRIVER occurrences can be deleted when the corresponding COVER occurrence is no longer required.

Sample Documentation

DATA MODEL : Motor Insurance Application

Page 20

Author : Patric Downes

Date : 3 February 1989 1989

ENTITY TYPE : 17 - DRIVER, continued

Relationship Population:

> On average, each COVER occurrence will have 1.5 related drivers.

Source

Initial: Current data from Insurance Admin.

History from archive.

Ongoing: Clerical input as a result of New Business, Adjustments, Renewals.

Questions

1. If more than one VEHICLE is covered, do the DRIVERS depend on the specific VEHICLE?

2. This data structure results in new DRIVER occurrence being created each time COVER is altered. Is this broadly satisfactory or should DRIVERS be independent of, although related to, the COVER?

DATA MODEL : Motor Insurance Application   Page 24

Author : Patric Downes

Date : 15 February 1989

ENTITY TYPE :   20 - CUSTOMER

Description

The CUSTOMER entity type is used to hold details not just of policyholders but additionally any other interested party in the POLICY, eg HP company, Leasing Agent etc.

Attributes

CUSTOMER ID	CUSTOMER NAME	DATE OF BIRTH	SEX	OCCUPATION
I				
100243	Jane Marsden	09-10-54	F	Teacher
466748	Jon Beven	19-04-53	M	Line Manager
923611	International HP	-	-	-
800127	David Hunt	25-03-41	M	Administrator
714611	Brian Legge	16-11-60	M	Labourer

Hypothetical, Illustrative.

Volumetrics

Initial: Most POLICIES have only a single CUSTOMER. It is rare for a CUSTOMER to have more that 1 POLICY. About 10% of POLICIES will have additional 'customers' eg. HP companies etc. For current volumes, including history, there will be approx. <u>33,000</u> CUSTOMERS.

Growth: As for POLICY + 10% ie. about <u>5,500</u> per annum.

Retention: Delete with last related POLICY.

Relationship Population:

        Not applicable.

Sample Documentation   363

DATA MODEL : Motor Insurance Application

Author : Patric Downes

Date : 15 February 1989

ENTITY TYPE : 20 - CUSTOMER, continued

Source

Initial: Admin records.

Ongoing: New business, Adjustment and Renewals sections.

Questions

1. Is there advantage in having a CUSTOMER TYPE to differentiate between real CUSTOMERS and other interested parties?

DATA MODEL : Motor Insurance Application   Page 26

Author : Patric Downes

Date : 19 February 1989

ENTITY TYPE : 21 - ADDRESS

Description

The entity type contains address data for each CUSTOMER. The entity type has both an effective date ('From Date') and an end date. The end date will not be completed for current addresses.

Attributes

CUSTOMER ID	FROM DATE	END DATE	ADDRESS LINE 1	ADDRESS LINE 6	POST CODE
I/S	I				
466748	01-10-84	19-04-86	6 Church Road	Essex	CM12 6QT
466748	20-04-86		The Cottage	Herts	JP41 9WX
923611	15-01-78		New House	Wales	CD1 1AA
800127	17-11-84		16 The Leys	Lancs	WN5 9XA
714611	15-02-79		25 High Street	York	LS21 9UA

Hypothetical, Illustrative.

Volumetrics

Initial: Approx 5% CUSTOMERS have more than one ADDRESS. The volume is therefore 33,000 + 5% = <u>34,650</u>

Growth: As for CUSTOMER, Approx 5,500 perannum.

Retention: Delete ADDRESS two years after End Date or with CUSTOMER.

Relationship Population:
                To CUSTOMER: Few CUSTOMERS have more than 1 ADDRESS.

Sample Documentation   365

DATA MODEL : Motor Insurance Application   Page 27

Author : Patric Downes

Date : 19 February 1989

ENTITY TYPE : 21 - ADDRESS, continued

Source

Initial: Admin records.
Ongoing: New Business, Adjustments and Renewals sections.

Questions

None.

DATA MODEL : Motor Insurance Application   Page   28

Author : Patric Downes

Date : 24 February 1989

ENTITY TYPE :   22 - AREA CODE

Description

The AREA CODE is an insurer specific code that relates to the risk factor perceived for that area. Areas are defined and perceived differently by each INSURER. Each AREA CODE has an effective date and relates to a single INSURER.

Attributes

INSURER CODE	AREA CODE	START DATE	END DATE	DESCRIPTION
I/S	I	I		
UK	A	05-04-88	-	London Postal
UK	B	05-04-88	-	Greater London, Glasgow, Liverpool
UK	C	05-04-88	-	Manchester, Birmingham, Edinburgh, Cardiff
UK	D	05-04-88	-	Other Urban Areas
UK	E	05-04-88	-	Rural Areas
UK	F	05-04-88	-	Isle of Man, Isle of Wight

Hypothetical, Illustrative

Volumetrics

Initial: No history is required. INSURERS have an average of six area codes. The initial volume is therefore about <u>60</u>.

Growth: Insurers reassess area codes about once every four years. Therefore growth will be 60 x 25% + 1 new scheme = <u>20/25</u> per annum.

Retention: Retain for two years after expiry.

Sample Documentation   367

DATA MODEL : Motor Insurance Application

Author : Patric Downes

Date : 24 February 1989

ENTITY TYPE : 22 - AREA CODE, continued

Relationship Population:

    To INSURER: Allowing for build up of history, about 12 AREA CODES per INSURER.

Source

Initial: Insurance Admin.

Ongoing: Brokerage Control.

Questions

None.

DATA MODEL : Motor Insurance Application

Page 30

Author : Patric Downes

Date : 27 February 1989

ENTITY TYPE :   23 - ADDRESS IN AREA

Description

This entity type provides a cross-reference link between the INSURER specific AREA CODES and actual ADDRESSES of CUSTOMERS. This enables the right AREA CODE to be used for premium calculation and risk assessment purposes. Each ADDRESS IN AREA occurrence is related to one AREA CODE and one ADDRESS.

Attributes

CUSTOMER ID	ADDRESS START DATE	INSURER CODE	AREA CODE	AREA CODE START DATE
I/S	I/S	I/S	I/S	I/S
466748	20-04-86	UK	D	05-04-88
923611	15-01-78	UK	C	05-04-88
800127	17-11-84	UK	D	05-04-88

Hypothetical, Illustrative

Volumetrics

Initial: Due to the low level of change in AREA CODES, there is likely to be only one associated AREA CODE for each ADDRESS held. The initial volume is therefore <u>34,650</u>.

Growth: Linked to the growth rate of ADDRESSES. Approx. <u>5,500</u> per annum.

Retention: Delete along with associated ADDRESSES.

Relationship Population:

    To ADDRESS: Normally only one ADDRESS IN AREA occurrence.

    To AREA CODE: Given a normal distribution across areas, there will be approx 580 occurrences for each AREA CODE.

Sample Documentation

DATA MODEL : Motor Insurance Application

Author : Patric Downes

Date : 27 February 1989

ENTITY TYPE : 23 - ADDRESS IN AREA, continued

Source

Initial: Insurance Admin.

Ongoing: Created as new addresses are added ie. New Business, Adjustments and Renewals.

Questions

1. Should the AREA CODE be embedded into the ADDRESS or is the need to preserve history important?

DATA MODEL : Motor Insurance Application

Author : Patric Downes

Date : 1 March 1989

ENTITY TYPE : 24 - INTEREST TYPE

Description

This entity type defines the different ways in which a CUSTOMER can be related to a POLICY.

Attributes

INTEREST TYPE	DESCRIPTION
I	
C	Current Policyholder
P	Prior Policyholder
H	Hire-Purchase Company
L	Leasing Company

Real, Exhaustive

Volumetrics

Initial:	As shown above. Currently 4.
Growth:	Minimal.
Retention:	Infinite.
Relationship Population:	N/A.

Source

Initial:	Insurance Admin.
Ongoing:	Insurance Admin.

Questions

None.

DATA MODEL : Motor Insurance Application   Page 33

Author :   Patric Downes

Date :   4 March 1989

ENTITY TYPE :   25 - INTEREST IN POLICY

Description

This entity type identifies the exact relationship between a POLICY and one or more CUSTOMERS. It also allows a CUSTOMER to have an interest in several POLICIES. Each INTEREST IN POLICY occurrence is related to one POLICY, and is categorised by a specific INTEREST TYPE.

Attributes

POLICY NO.	CUSTOMER ID.	INTEREST TYPE CODE
I/S	I/S	I/S
MP10116	466748	C
MP10116	800127	P
MP10116	923611	H

Hypothetical, Illustrative

Volumetrics

Initial: Only approx 10% of POLICIES will have more than one related CUSTOMER. The estimated volume is therefore: <u>33,000</u>.

Growth: As for POLICY plus 10% re. <u>5,500</u> per annum.

Retention: Delete with associated POLICY occurrence.

Relationship Population:

　　　　To POLICY: Normally one per POLICY, 10% have 2.

　　　　To CUSTOMER: Normally one per CUSTOMER.

Practical Data Analysis

DATA MODEL : Motor Insurance Application   Page 34

Author : Patric Downes

Date : 4 March 1989

ENTITY TYPE : 25 - INTEREST IN POLICY, continued

Relationship Population - continued

> To INTEREST TYPE: Most occurrences will be to Interest Type Code C - Current policyholder. About 10 per cent of occurrences will be distributed across the other codes, ie. about 1000 each.

Source

Initial: Insurance Admin.

Ongoing: New Business Adjustments and Renewals sections.

Questions

1. Is a history required of INTEREST IN POLICY?

2. Should the occurrence carry an effective date?

PROCESS ANALYSIS DOCUMENTATION

The Process Analysis documentation shown here concentrates on the 'Incept New Policy' Process used in Part 3 of the book and applies this to the data model described in this Appendix. It contains the following items:

- An Overall Introduction to the Process;

- List of Operations with Start and End Events;

- Flow of Operations;

- Descriptions of each operation;

- Process Maps for each operation;

- Cross-reference charts to entity types and relationships;

- A 'CRUD' chart for the whole process.

It should be noted that as the data model itself is incomplete, then the processes are difficult to define and to map. For example, the processes recognise that documentation has to be issued but the data model has no means of showing this. This is a clear indication that further work is required and shows the value of analysing the area from both the data and process view points.

The operations shown differ slightly from those given in Part 3. The examples used here can be viewed as a later iteration in the analysis. For example, operations 1 and 2 from Figure 12.4 have been combined as one always follows the other so there is no point in differentiating between them.

DATA MODEL : Motor Insurance Application

Author : Patric Downes

Date : 15 January 1989

PROCESS: Incept New Policy
OVERALL DESCRIPTION

The Incept New Policy process is concerned with the initial creation and administration of a policy. It begins with receipt of a proposal form and, usually, payment based on a prior quotation. The policy and customer details are set up and the details checked. A number of actions then follow. If all details are in order the premium is calculated and any adjustments against the amount received are outlined. If detail is incomplete then the customer is asked to supply additional data.

Policy documents are only issued when all details and sufficient funds have been received, otherwise a cover note is issued and diarised for follow-up.

DATA MODEL : Motor Insurance Application

Page: 2

Author : Patric Downes

Date : 17 January 1989

PROCESS : Incept New Policy
OPERATIONS : LIST OF OPERATIONS

START EVENT	OPERATION	END EVENT
1. Proposal Received	Set up customer address and risk data. Check if complete.	Customer and policy created.
2. If risk data missing	Request additional data.	Letter issued.
3. All data complete	Calculate Premium and compare to received amount.	Set up payment due for policy.
4. Overpaid policy	Issue Refund.	Cheque dispatched.
5. Payments in balance	Issue policy documents and post to underwriter.	Policy complete.
6. Underpaid policy	Issue payment request.	Payment slip despatched.
7. Waiting customer action	Issue cover note and diarise.	Cover note issued.
8. Receive additional data	Check and log additional data.	Policy updated.

DATA MODEL : Motor Insurance Application

Page: 3

Author : Patric Downes

Date : 17 January 1989

PROCESS : Incept New Policy

OPERATIONS : FLOW OF OPERATIONS

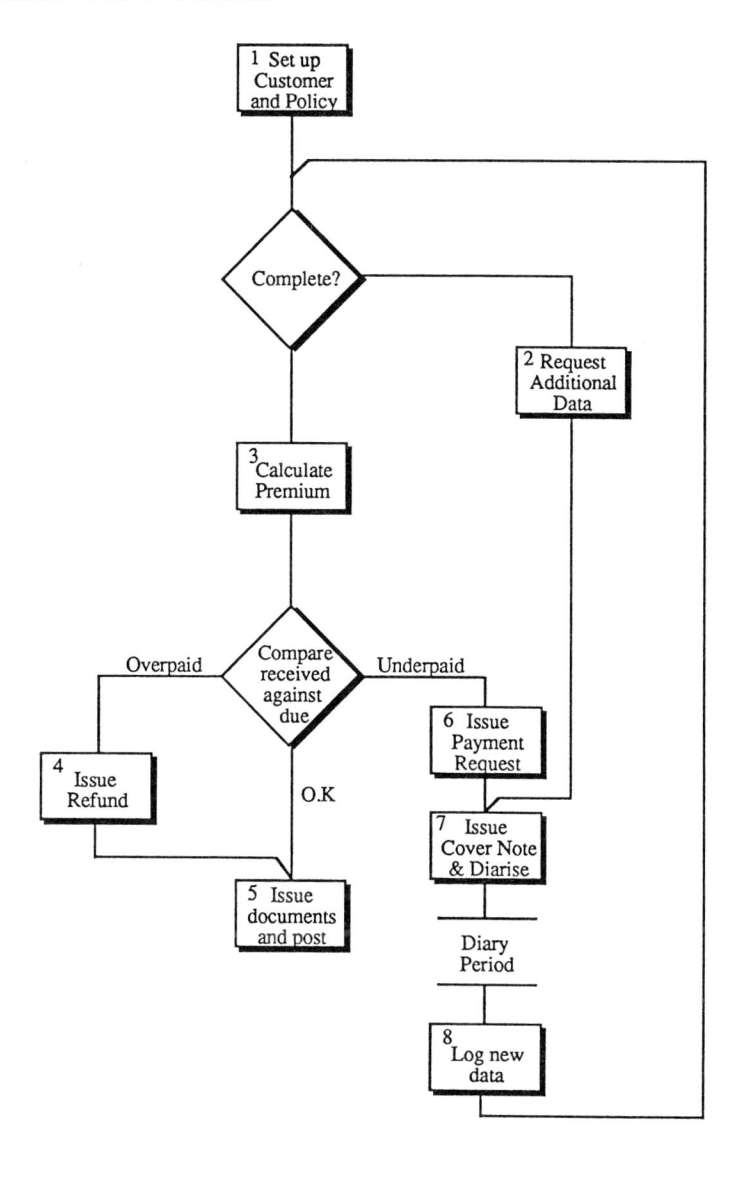

```
                    ┌─────────────┐
                    │ 1 Set up    │
                    │  Customer   │
                    │  and Policy │
                    └──────┬──────┘
                           │
                        ◇ Complete? ─── No ──► ┌─────────────┐
                           │                   │ 2 Request   │
                          Yes                  │ Additional  │
                           │                   │    Data     │
                    ┌──────▼──────┐            └─────────────┘
                    │ 3 Calculate │
                    │   Premium   │
                    └──────┬──────┘
                           │
              Overpaid  ◇ Compare  Underpaid
           ┌──────────── received ──────────┐
           │            against due         │
           │               │                │
           │              O.K.      ┌───────▼───────┐
           │               │        │ 6 Issue       │
    ┌──────▼──────┐        │        │   Payment     │
    │ 4 Issue     │        │        │   Request     │
    │   Refund    │        │        └───────┬───────┘
    └──────┬──────┘        │                │
           │               │        ┌───────▼───────┐
           │               │        │ 7 Issue       │
           │               │        │ Cover Note    │
           │               │        │ & Diarise     │
           │               │        └───────┬───────┘
           │       ┌───────▼───────┐        │
           └──────►│ 5 Issue       │    Diary Period
                   │  documents    │        │
                   │  and post     │        │
                   └───────────────┘        │
                                    ┌───────▼───────┐
                                    │ 8 Log new     │
                                    │    data       │
                                    └───────────────┘
```

Practical Data Analysis

379

DATA MODEL : Motor Insurance Application   Page: 4

Author : Patric Downes

Date : 21 January 1989

PROCESS: Incept New Policy
OPERATIONS : OPERATION 1: SET UP CUSTOMER/POLICY

Description: This operation creates a new policy and if necessary creates new customer and vehicle details. The cover details are set up with data from the proposal and any payment received is logged. The details are checked for completeness.

Responsibility: New Business Clerks.

Events: Start: Receipt of proposal.

End: Customer and policy created.

Frequency: Approximately 5,000 per annum.

Response Time: About 5 minutes in total.

Criticality: It is important to process all proposals on day of receipt to prevent backlogs. Cash should be logged, reconciled and banked by 3pm.

DATA MODEL : Motor Insurance Application   Page: 5

Author : Patric Downes

Date : 21 January 1989

PROCESS: Incept New Policy

OPERATIONS : OPERATION 1: SET UP CUSTOMER/POLICY - PROCESS MAP

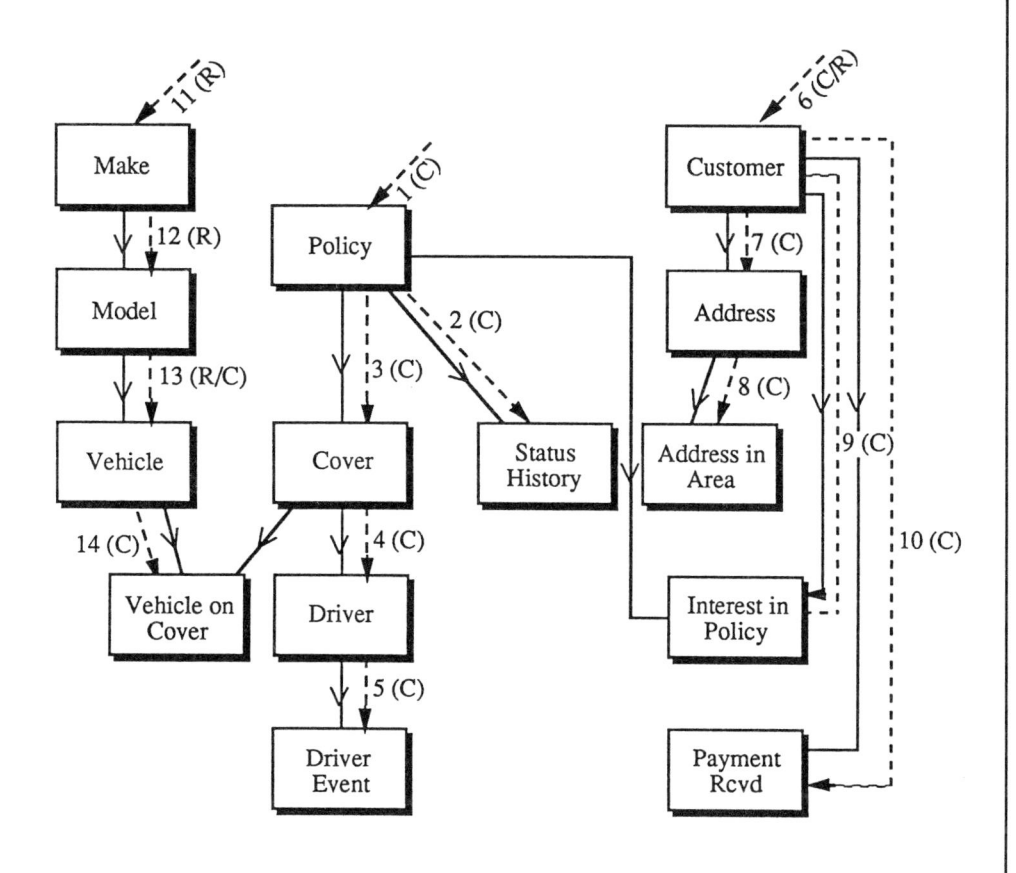

Practical Data Analysis   381

DATA MODEL : Motor Insurance Application   Page: 6

Author : Patric Downes

Date : 17 January 1989

PROCESS: Incept New Policy
OPERATIONS: OPERATION 2: REQUEST ADDITIONAL DATA

Description: Where policy details are missing or unclear, then a letter has to be sent to the customer to request the missing data. This can be a standard letter for most situations or may be non-standard for unusual circumstances. A summary of the letter is recorded with its originator and date.

Responsibility: Standard: New Business Clerk.

Non Standard: Section Head.

Events: Start: Missing data noted.

End: Letter created.

Frequency: About 10 per cent of proposal, ie. 500 per annum.

Response Time: 2 minutes for simple letters, 15 minutes for complex.

Criticality: Rapid responses to queries is beneficial. Should be achieved on same day as missing data noted.

DATA MODEL : Motor Insurance Application

Page: 7

Author : Patric Downes

Date : 17 January 1989

PROCESS: Incept New Policy

OPERATIONS : OPERATION 2: REQUEST ADDITIONAL DATA - PROCESS MAP

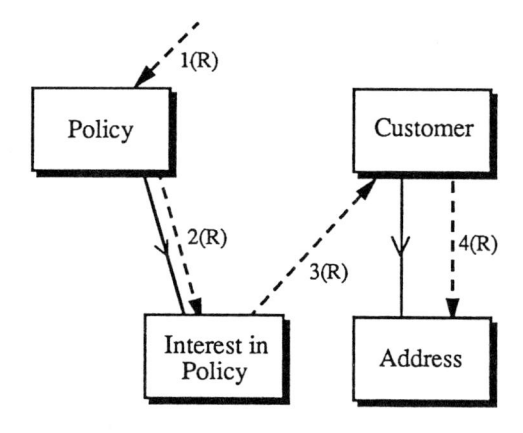

N.B. This map is incomplete. It is possible to obtain the address, but there is no means of storing letter details

Practical Data Analysis 383

DATA MODEL : Motor Insurance Application

Page: 8

Author : Patric Downes

Date : 27 January 1989

PROCESS: Incept New Policy

OPERATIONS : OPERATION 3: CALCULATE PREMIUM

Description: The required premium is calculated based on the policy and cover details using the appropriate insurer's rates. A summary of the calculation is stored, and a payment due set up for the calculated amount.

Responsibility: New Business Clerks.

Events: Start: Complete policy detail.

End: Payment due created.

Frequency: Once for each new policy, ie. 5,000 per annum.

Response Time: Manually this can take 10 minutes. With computer assistance this should be less than 5 seconds.

Criticality: Accurate and rapid premium calculation is important. Errors can be very costly.

DATA MODEL : Motor Insurance Application   Page: 9

Author : Patric Downes

Date : 27 January 1989

PROCESS: Incept New Policy
OPERATIONS: OPERATION 3: CALCULATE PREMIUM - PROCESS MAP

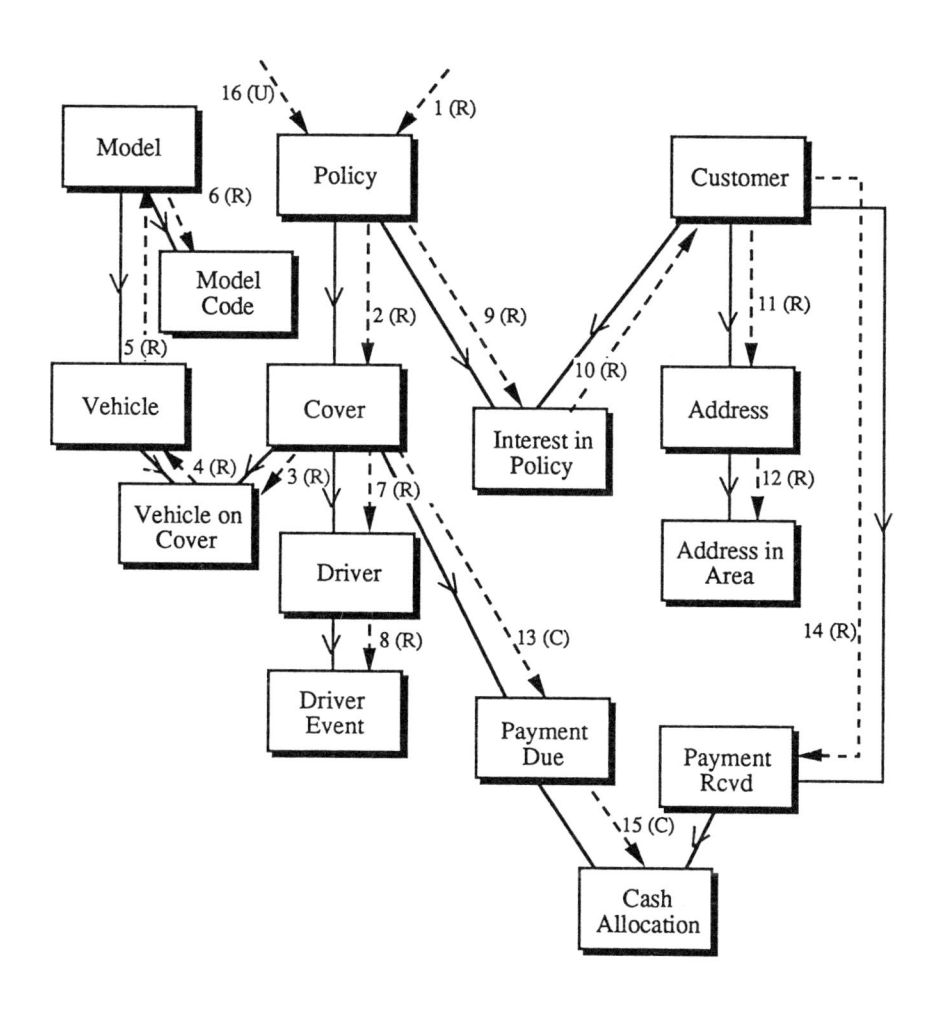

Practical Data Analysis   385

DATA MODEL : Motor Insurance Application

Page: 10

Author : Patric Downes

Date : 27 January 1989

PROCESS: Incept New Policy

OPERATIONS: OPERATION 4: ISSUE REFUND

Description: Policies are occasionally overpaid on inception. When the premium due is less than that received then a refund cheque has to be issued.

Responsibility: Issue: New Business Clerks.

Authorisation: Up to £15 - Clerk

Up to £50 - Section Head

Over £50 - Department Manager.

Events: Start: Overpaid policy

End: Cheque issued.

Frequency: Approx 100 per annum.

Response Time: Variable on authorisation needed.

Criticality: Not time critical.

DATA MODEL : Motor Insurance Application

Page: 11

Author : Patric Downes

Date : 27 January 1989

PROCESS: Incept New Policy

OPERATIONS: OPERATION 4: ISSUE REFUND - PROCESS MAP

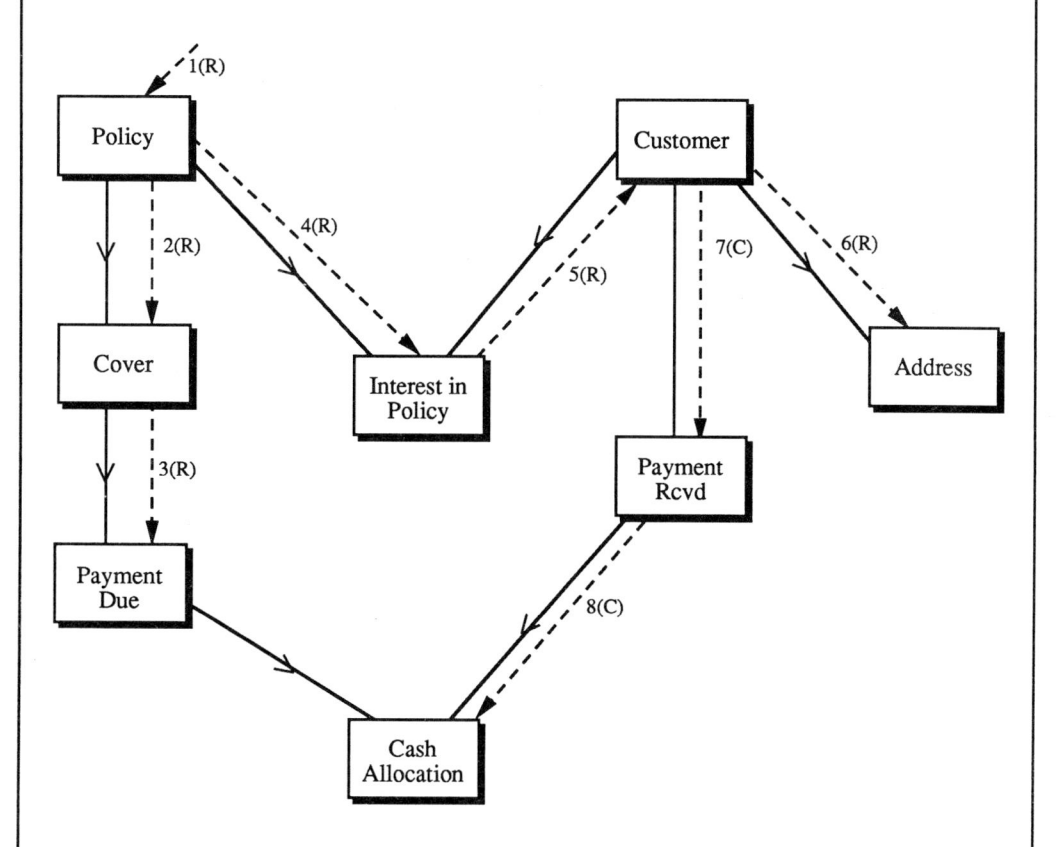

```
         1(R)
    ┌────────┐                          ┌──────────┐
    │ Policy │                          │ Customer │
    └────────┘                          └──────────┘
         │  ╲ 4(R)                      ╱    │     ╲ 6(R)
         │ 2(R)    ╲                 ╱  5(R) │ 7(C)  ╲
         ▼              ╲         ╱          │         ▼
    ┌────────┐           ┌──────────────┐    │    ┌──────────┐
    │ Cover  │           │  Interest in │    │    │ Address  │
    └────────┘           │   Policy     │    │    └──────────┘
         │               └──────────────┘    ▼
         │ 3(R)                          ┌──────────┐
         ▼                               │ Payment  │
    ┌────────┐                           │  Rcvd    │
    │Payment │                           └──────────┘
    │  Due   │                      ╱   8(C)
    └────────┘               ╱
              ╲         ╱
               ┌──────────────┐
               │    Cash      │
               │  Allocation  │
               └──────────────┘
```

N.B This process is probably incomplete as lack of a 'correspondence' entity type prohibits storage of despatch details

DATA MODEL : Motor Insurance Application

Author : Patric Downes

Date : 28 January 1989

PROCESS: Incept New Policy
OPERATIONS: OPERATION 5: ISSUE DOCUMENTS/POST

Description: When the policy payments are in balance and all details are available, the formal policy documents can be issued to the customer and the policy 'posted' (notified) to the insurer.

Responsibility: New Business Clerks.

Events: Start: Policy details complete **and** premium in balance.
End: Incepted policy.

Frequency: Few policies are abandoned before inception, so 5,000 per annum.

Response Time: Less than 5 minutes.

Criticality: Insurers should be notified within 48 hours of policy inception.

DATA MODEL : Motor Insurance Application   Page: 13

Author : Patric Downes

Date : 28 January 1989

PROCESS: Incept New Policy

OPERATIONS : OPERATION 5: ISSUE DOCUMENTS - PROCESS MAP

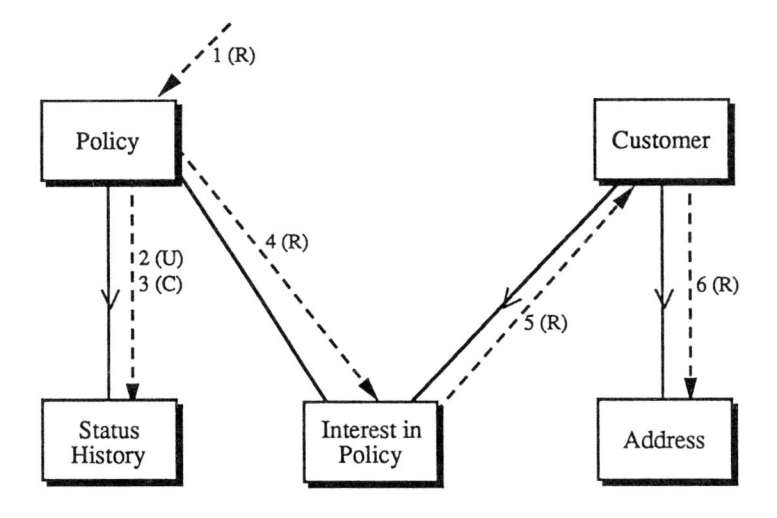

```
                1 (R)
                  ↘
    ┌─────────┐                              ┌──────────┐
    │ Policy  │                              │ Customer │
    └─────────┘                              └──────────┘
       │   ╲       4 (R)                          ↑ │
       │2(U)  ╲                                   │ │ 6 (R)
       │3(C)    ╲                            5(R) │ │
       ↓          ↘                               │ ↓
    ┌─────────┐     ┌────────────┐           ┌─────────┐
    │ Status  │     │ Interest in│           │ Address │
    │ History │     │  Policy    │           │         │
    └─────────┘     └────────────┘           └─────────┘
```

N.B. Incomplete process map. No means of storing details of issued documents

Practical Data Analysis   389

DATA MODEL : Motor Insurance Application

Page: 14

Author : Patric Downes

Date : 30 January 1989

PROCESS: Incept New Policy

OPERATIONS : OPERATION 6: ISSUE PAYMENT REQUEST

Description: In some cases, the amount paid by the customer will not equal the calculated premium. A payment request should be sent for the difference in value.

Responsibility: New Business Clerks.

Events: Start: Policy details complete <u>and</u> underpaid policy.

End: Payment request despatched.

Frequency: About 15% of policies, ie. 750 per annum.

Response Time: Within 5 minutes of premium calculation.

Criticality: Payment requests should be issued daily.

DATA MODEL : Motor Insurance Application

Page: 15

Author : Patric Downes

Date : 30 January 1989

PROCESS: Incept New Policy

OPERATIONS : OPERATION 6: ISSUE PAYMENT REQUEST - PROCESS MAP

Mapping not possible until further analysis in carried out.

DATA MODEL : Motor Insurance Application

Page: 16

Author : Patric Downes

Date : 17 January 1989

PROCESS: Incept New Policy

OPERATIONS: OPERATION 7: ISSUE COVER NOTE

Description: Cover notes are issued where a policy cannot be completely processed. Details of the cover note should be logged and a diary entry set up to ensure that outstandings are progressed.

Responsibility: New Business Clerks.

Events: Start: Payment Request Issued;

Letter Issued.

End: Cover note issued.

Frequency: Approx 1,250 cover notes issued per annum.

Response Time: About 2 minutes.

Criticality: Cover notes must be issued immediately if needed to ensure the customer is not in breach of the law.

DATA MODEL : Motor Insurance Application Page: 17

Author : Patric Downes

Date : 17 January 1989

PROCESS: Incept New Policy
OPERATIONS : OPERATION 7: ISSUE COVER NOTE - PROCESS MAP

Mapping not yet possible.

DATA MODEL : Motor Insurance Application   Page: 18

Author : Patric Downes

Date : 3 February 1989

PROCESS: Incept New Policy
OPERATIONS : OPERATION 8: LOG NEW DATA

Description: Any additional data received from the customer is logged, and records updated.

Responsibility: New Business Clerks.

Events: Start: Diary Prompt Expiry;
                          Customer communication received.
                   End: Inception reactivated.

Frequency: About 750 per annum.

Response Time: All post and diary prompts should have been reactivated by 10am.

Criticality: Quality of customer service requires daily response to letters etc.

DATA MODEL : Motor Insurance Application   Page: 19

Author : Patric Downes

Date : 3 February 1989

PROCESS: Incept New Policy

OPERATIONS : OPERATION 8: LOG NEW DATA - PROCESS MAP

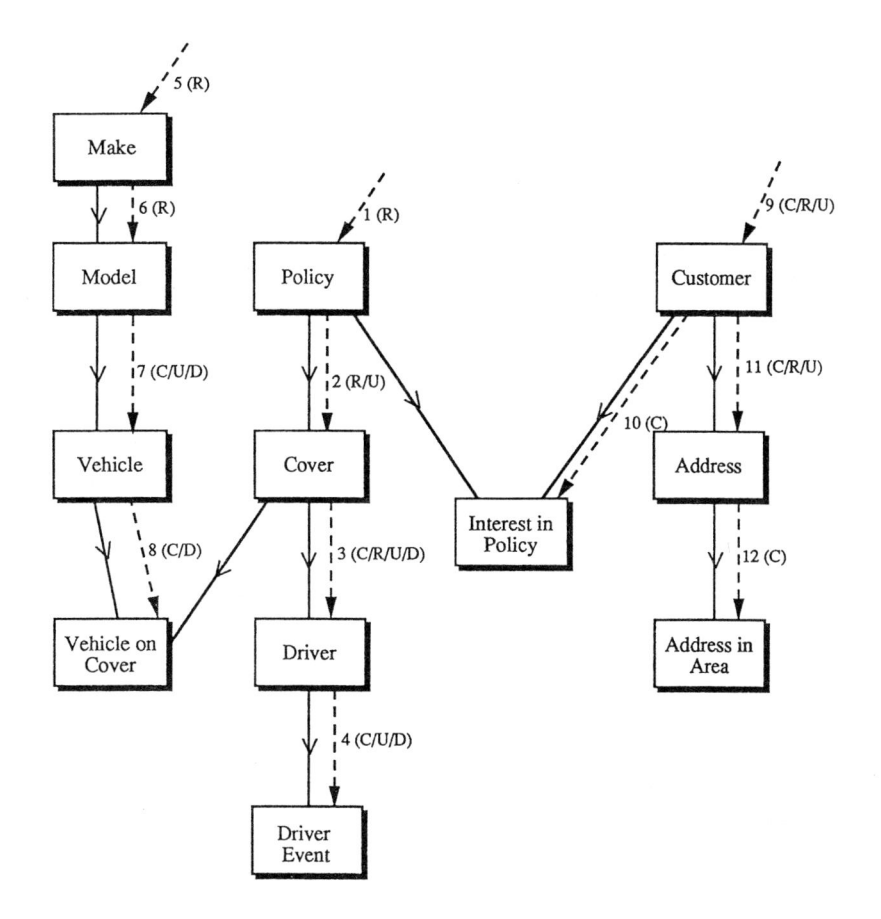

```
        5 (R)
         ↙
      ┌──────┐
      │ Make │
      └──────┘
         │ 6 (R)
         ↓                    1 (R)                         9 (C/R/U)
      ┌──────┐           ┌────────┐                       ┌──────────┐
      │Model │           │ Policy │                       │ Customer │
      └──────┘           └────────┘                       └──────────┘
         │ 7 (C/U/D)         │ 2 (R/U)                        │ 11 (C/R/U)
         ↓                   ↓                                ↓
      ┌───────┐         ┌───────┐       10 (C)           ┌─────────┐
      │Vehicle│         │ Cover │                         │ Address │
      └───────┘         └───────┘                         └─────────┘
          │ 8 (C/D)        │  3 (C/R/U/D)    ┌──────────┐    │ 12 (C)
          ↓                ↓                 │Interest in│    ↓
      ┌───────┐        ┌────────┐            │  Policy   │  ┌──────────┐
      │Vehicle│        │ Driver │            └──────────┘   │Address in│
      │on Cover│       └────────┘                           │   Area   │
      └───────┘            │ 4 (C/U/D)                      └──────────┘
                           ↓
                       ┌────────┐
                       │ Driver │
                       │ Event  │
                       └────────┘
```

N.B. There is considerable variability of action in this operation as the actions will depend on the quality of data supplied e.g. for alterations to driver details, the new data may cause

- New driver to be created (C)
- Amended driver events (R)
- Updated driver details (U)
- Drivers to be removed (D)

or any combination of these

Practical Data Analysis   395

DATA MODEL : Motor Insurance Application   Page: 20

Author : Patric Downes

Date : 3 February 1989

PROCESS: Incept New Policy
OPERATIONS : PROCESS\ENTITY TYPE CROSS-REFERENCE

ENTITY TYPE	Set up Customer/ Policy	Request Additional Data	Calculate Premium	Issue Refund	Issue Documents	Payment Request	Issue Cover Note	Log New Data
Make	√							√
Vehicle Type								
Model	√		√					√
Vehicle	√		√					√
Vehicle Modification								
Insurer								
Vehicle Code								
Model Code			√					
Policy Type								
Policy	√	√	√	√	√			√
Special Information								
Policy Status								
Status History	√				√			
Cover Type								
Cover	√		√	√				√
Vehicle on cover	√		√					√
Driver	√		√					√
Event Code								
Driver Event	√		√					√
Customer	√	√	√	√	√			√
Address	√	√	√	√	√			√
Area Code								
Address in Area	√		√					√
Interest Type								
Interest in Policy	√	√	√	√	√			√
Payment Due			√	√				
Payment Rcvd	√		√	√				
Cash Allocation			√	√				

396   Sample Documentation

DATA MODEL : Motor Insurance Application   Page: 21

Author : Patric Downes

Date : 3 February 1989

PROCESS: Incept New Policy

OPERATIONS : RELATIONSHIP/PROCESS CROSS-REFERENCE

Relationship	Set up Customer/ Policy	Request Additional Data	Calculate Premium	Issue Refund	Issue Documents	Payment Request	Issue Cover Note	Log New Data
Make - Model	√							√
Vehicle Type - Model								
Model - Vehicle	√		√					√
Model - Model Code			√					
Vehicle - Vehicle Modification								
Vehicle - Vehicle on Cover	√		√					√
Insurer - Vehicle Code								
Insurer Policy								
Insurer - Area Code								
Vehicle Code - Model Code								
Policy Type - Policy								
Policy - Special Information								
Policy - Status History					√			
Policy - Cover	√		√	√				√
Policy-Interest in Policy		√	√	√	√			
Policy Status - Status History								
Cover Type - Cover								
Cover - Vehicle on Cover			√					
Cover - Driver	√		√					√
Cover - Payment Due			√	√				
Driver - Driver Event	√		√					√
Event Code - Driver								
Customer - Address	√	√	√	√	√			√
Customer - Interest in Policy	√	√	√	√	√			√
Customer - Payment Received	√		√	√				
Address - Address in Area			√					√
Area Code - Address in Area	√							
Interest Type - Interest in Policy								
Payment Due - Cash Allocation			√					
Pay Received - Cash Allocation				√				

DATA MODEL : Motor Insurance Application

Page: 22

Author : Patric Downes

Date : 3 February 1989

PROCESS: Incept New Policy

OPERATIONS : "CRUD" CHART

Entity Type	Set up Customer/ Policy	Request Additional Data	Calculate Premium	Issue Refund	Issue Documents	Payment Request	Issue Cover Note	Log New Data
Make	R							R
Vehicle Type								
Model	R		R					R
Vehicle	CR		R					CUD
Vehicle Modification								
Insurer								
Vehicle Code								
Model Code			R					
Policy Type								
Policy	C	R	RU	R	R			R
Special Information								
Policy Status								
Status History	C				UC			
Cover Type								
Cover	C		R	R				RU
Vehicle on cover	C		R					CD
Driver	C		R					CRUD
Event Code								
Driver Event	C		R					CUD
Customer	CR	R	R	R	R			CRU
Address	C	R	R	R	R			CRU
Area Code								
Address in Area	C		R					C
Interest Type								
Interest in Policy	C	R	R	R	R			C
Payment Due			C	R				
Payment Rcvd	C		R	C				
Cash Allocation			C	C				

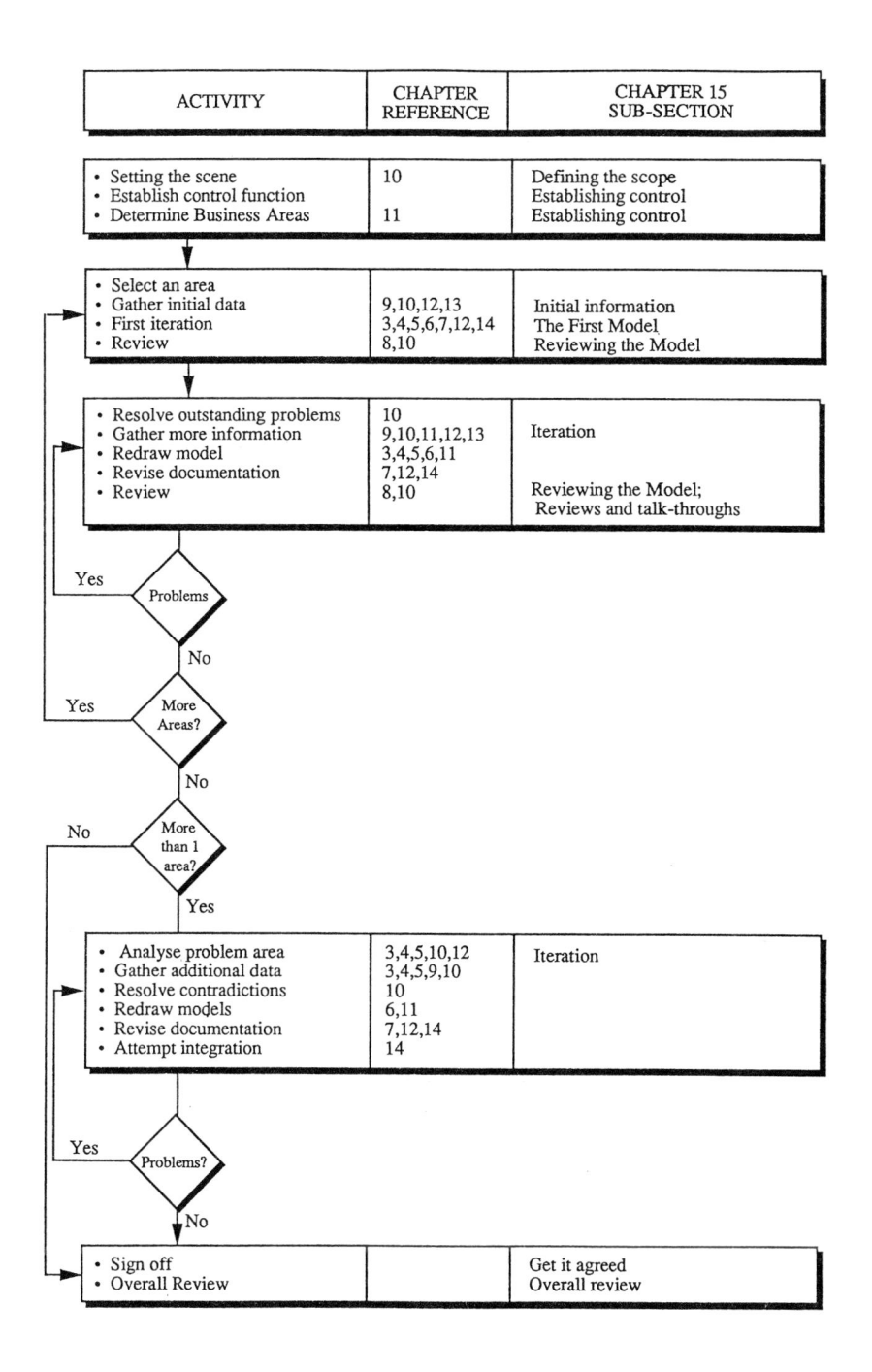

ACTIVITY	CHAPTER REFERENCE	CHAPTER 15 SUB-SECTION
• Setting the scene • Establish control function • Determine Business Areas	10 11	Defining the scope Establishing control Establishing control
• Select an area • Gather initial data • First iteration • Review	 9,10,12,13 3,4,5,6,7,12,14 8,10	 Initial information The First Model Reviewing the Model
• Resolve outstanding problems • Gather more information • Redraw model • Revise documentation • Review	10 9,10,11,12,13 3,4,5,6,11 7,12,14 8,10	Iteration Reviewing the Model; Reviews and talk-throughs

Yes — Problems — No
Yes — More Areas? — No
No — More than 1 area? — Yes

| • Analyse problem area
• Gather additional data
• Resolve contradictions
• Redraw models
• Revise documentation
• Attempt integration | 3,4,5,10,12
3,4,5,9,10
10
6,11
7,12,14
14 | Iteration |

Yes — Problems? — No

| • Sign off
• Overall Review | | Get it agreed
Overall review |

METHODOLOGY CROSS-REFERENCE